Edward Tyson Reichert, Silas Weir Mitchell

Researches upon the Venoms of Poisonous Serpents

Edward Tyson Reichert, Silas Weir Mitchell

Researches upon the Venoms of Poisonous Serpents

ISBN/EAN: 9783337034856

Printed in Europe, USA, Canada, Australia, Japan

Cover: Foto ©berggeist007 / pixelio.de

More available books at **www.hansebooks.com**

SMITHSONIAN CONTRIBUTIONS TO KNOWLEDGE.
647

RESEARCHES

UPON THE

VENOMS OF POISONOUS SERPENTS.

BY

S. WEIR MITCHELL, M.D.,
MEMBER OF THE NATIONAL ACADEMY OF SCIENCES, U. S. A.; PRESIDENT OF THE COLLEGE OF PHYSICIANS
OF PHILADELPHIA.

AND

EDWARD T. REICHERT, M.D.,
PROFESSOR OF PHYSIOLOGY IN THE UNIVERSITY OF PENNSYLVANIA.

[ACCEPTED FOR PUBLICATION, MAY, 1885.]

WASHINGTON:
PUBLISHED BY THE SMITHSONIAN INSTITUTION
1886.

COMMISSION

TO WHICH THIS MEMOIR HAS BEEN REFERRED.

JOHN S. BILLINGS, M. D.,
HENRY G. BEYER, M. D.

SPENCER F. BAIRD,
Secretary S. I.

COLLINS PRINTING HOUSE,
PHILADELPHIA.

PREFACE.

The authors desire to express their multiple obligations to the Smithsonian Institution. They have to thank the Army Medical Library for the valuable Bibliography appended to this essay. With that to be found in Dr. Weir Mitchell's former essay, it completes the list of such knowledge up to January, 1885. They desire also to thank Her Britannic Majesty's Indian Government for help in securing Indian serpent poisons. Among individuals they owe to no one so deep a debt as to Vincent Richards, Esq., of Goalundo, British India. Without his untiring aid the authors feel that it would have been impossible to have extended their inquiries beyond our native snakes.

The excellent plates were drawn for the most part by Dr. J. Madison Taylor, and thanks are due to Dr. Geo. A. Piersol's skill for the interesting microphotographs of blood-corpuscles attacked by venom. The authors are also indebted to Dr. Guy Hinsdale for having made the tabulated reductions of kymographion tracings.

S. WEIR MITCHELL,
EDWARD T. REICHERT.

Physiological Laboratory of the
University of Pennsylvania.

TABLE OF CONTENTS.

	PAGE
PREFACE	iii
LIST OF ILLUSTRATIONS	ix
INTRODUCTION	1

CHAPTER I.
PHYSICAL CHARACTERISTICS OF VENOM . . . 5

CHAPTER II.
THE CHEMISTRY OF VENOMS . . . 9

CHAPTER III.
THE EFFECTS OF VARIOUS AGENTS ON VENOM 21

CHAPTER IV.
THE EFFECTS OF VENOM WHEN APPLIED TO MUCOUS OR SEROUS SURFACES . . . 44

CHAPTER V.
THE EFFECTS OF VENOM ON THE NERVOUS SYSTEM . 48

CHAPTER VI.
THE GLOBULINS AND PEPTONES COMPARED AS REGARDS LOCAL POISONOUS ACTIVITY . 51

CHAPTER VII.
THE ACTION OF VENOMS AND THEIR ISOLATED GLOBULINS AND PEPTONES UPON THE PULSE-RATE 56

Section I.—The action of pure venom upon the pulse-rate . . . 56
    The action of pure venom upon the pulse-rate of normal animals . . 57
    The action of pure venom upon the pulse-rate in animals with cut pneumogastric nerves 63
    The action of pure venom upon the pulse-rate of animals in which sections of the pneumogastric nerves and of the upper cervical portion of the spinal cord had been made 66

TABLE OF CONTENTS.

	PAGE
Section II.—The action of venom globulins upon the pulse-rate	69
The action of venom globulins upon the pulse-rate of normal animals	69
The action of venom globulins upon the pulse-rate of animals with cut pneumogastric nerves	74
The action of venom globulins upon the pulse-rate of animals with the pneumogastric nerves and cervical spinal cord cut	76
Section III.—The action of venom peptones upon the pulse-rate	79
The action of venom peptones upon the pulse-rate of normal animals	79
The action of venom peptones upon the pulse-rate of animals with cut pneumogastric nerves	82
The action of venom peptones upon the pulse-rate of animals with the pneumogastric nerves and cervical spinal cord cut	83

CHAPTER VIII.

THE ACTION OF VENOMS AND THEIR ISOLATED GLOBULINS AND PEPTONES UPON THE ARTERIAL PRESSURE 85

Section I.—The action of pure venom upon the arterial pressure	85
The action of pure venom upon the arterial pressure of normal animals	85
The action of pure venom upon the arterial pressure of animals with the pneumogastric nerves cut	92
The action of pure venom upon the arterial pressure of animals with the pneumogastric nerves and cervical spinal cord cut	95
The action of pure venom upon the arterial pressure of animals with the pneumogastric, depressor, and sympathetic nerves and spinal cord cut	98
Section II.—The action of venom globulins upon the arterial pressure	102
The action of venom globulins upon the arterial pressure of normal animals	102
The action of venom globulins upon the arterial pressure of animals with pneumogastric nerves cut	107
The action of venom globulins upon the arterial pressure of animals with pneumogastric, depressor, and sympathetic nerves and spinal cord cut	109
Section III.—The action of venom peptones upon the arterial pressure	112
The action of venom peptones upon the arterial pressure of normal animals	112
The action of venom peptones upon the arterial pressure of animals with the pneumogastric and depressor nerves cut	114
The action of venom peptones upon the arterial pressure of animals with the pneumogastric, depressor, and sympathetic nerves and cervical spinal cord cut	116

CHAPTER IX.

THE ACTION OF VENOMS AND THEIR ISOLATED GLOBULINS AND PEPTONES UPON RESPIRATION 119

Section I.—The action of pure venom upon the respiration	119
The action of pure venom upon the respiration of normal animals	119
The action of pure venom upon the respiration of animals with the pneumogastric nerves cut	122
Section II.—The action of venom globulins upon the respiration	125
The action of venom globulins upon the respiration of normal animals	125
The action of venom globulins upon the respiration of animals with the pneumogastric nerves cut	129

TABLE OF CONTENTS

	PAGE
Section III.—The action of venom peptones upon the respiration	130
The action of venom peptones upon the respiration of normal animals .	130
The action of venom peptones upon the respiration of animals with the pneumogastric nerves cut	131

CHAPTER X.

PATHOLOGY 133

CHAPTER XI.

GENERAL CONSIDERATIONS 153

BIBLIOGRAPHY 159
DESCRIPTION OF PLATES 181
INDEX 183

LIST OF WOOD-CUTS.

		PAGE
Figure 1.	Snake loop	3
Figure 2.	Venom dried	5
Figure 3.	Muscle tissue altered by venom	147
Figure 4.	Muscle tissue altered by venom	147

INTRODUCTION.

A FEW words of explanatory character in regard to the following essay may not be out of place. From the time of Fontana, 1767, until the able essay of Lucien Bonaparte, in 1843, on the chemistry of venom, there was no paper of moment on serpent poisons. In January, 1861, one of us, S. Weir Mitchell, published a long study of the venom of the *Crotalus durissus*, and in 1868 supplemented it by a shorter contribution, in which he related some recent discoveries of his own, and corrected certain errors of his former paper. These two essays may be considered as constituting with Lucien Bonaparte's the foundation of the later work in this direction, and perhaps as having left the study of venoms in as definite a position as could be gained with the laboratory facilities of 1843 to 1868.

In 1872, the government of India enabled Sir Joseph Fayrer to publish a volume of beautiful plates of the venomous snakes of India, to which was appended also a series of investigations into the toxicology of their poisons. In 1872 the same author and Dr. Lander Brunton contributed an admirable physiological study of the effects of venoms.[1]

In 1874, Vincent Richards, as chairman of a government commission, published an excellent report on antidotes.

Dr. Wall's[2] thoughtful and suggestive book appeared in 1883. It is a comparative study of the poisons of the colubrine and viperine serpents of India.

These, with a too brief study of the poison of our copperhead by Dr. Isaac Ott, of Easton, Pennsylvania, sum up all of value which has been added to the physiological literature of this most interesting subject.

Why it has won so few investigators is not far to seek. Even in India, where the appalling loss of life from snake-bites has of late invigorated research, the power and means of government were needed to overcome the obstacles which surround such scientific effort from inception to close. But, if in a land where snakes abound and professional snake-catchers can be had, it is yet not easy to follow this pursuit with success, elsewhere it is a task set about with inconceivable obstacles. The fear of serpents, the rarity of some species, the distances to which they have to be carried, the mortality of caged specimens, and the great cost of

---

[1] Proc. Roy. Soc. 1872, 1873, and 1875.
[2] Indian Snake Poisons; their Nature and Effects. A. J. Wall, M.D., F.R.Coll.S., 1883.
1 April, 1886.

purchase and transportation, need only to be mentioned as indicating our own difficulties. What had been done in India, sustained by a government, had to be with us attempted by private individuals, aided by the Smithsonian Institution, without which it would have been impossible to succeed. Our work began in the autumn of 1882, by extended efforts on our part, and that of the Smithsonian, to buy or otherwise get numerous living specimens of the American genera of Thanatophidiæ. This quest was kept up by every means our ingenuity could devise, and neither time nor money was spared. We succeeded in obtaining a sufficient number of rattlesnakes, including *Crotalus adamanteus* and *C. durissus*. We have had also enough of the Moccasin (*Ancistrodon piscivorus*). Our wants as regards Ground Rattlesnakes, Copperheads, and Coral-snakes have been less competently supplied, chiefly because these snakes are all small, so that to get enough of their poison for study it was essential to have a great many snakes. We have had in all about two hundred living serpents, and among them some superb specimens, which yielded poison in large quantities. Thus one—*C. adamanteus*—was eight and a half feet long and weighed nearly nineteen pounds. It furnished on one occasion about one and a half drachms of venom.

It was thought desirable by Prof. Baird and ourselves to examine the poisons of Indian serpents. To secure these the Secretary of State appealed to Her Majesty's Indian government in our behalf. A courteous response was returned, and orders given which resulted in our receiving a certain amount of Cobra venom. A more constant and larger supply was due to the generous and untiring kindness of Vincent Richards, Esq., M.R.C.S., of Goalundo, B. I.

The poison of the *Daboia Russellii*, the Indian viper, we sought in vain to secure. Government aid and private enterprise alike failed to secure a sufficient quantity of the venom of this dreaded reptile. The other Thanatophidiæ, of Australia, and South America, still await more careful study, and our preliminary report has already been the means of renewing interest in the chemical aspects of this study in India.

Such of our serpents as were not cared for by the hospitality of the Philadelphia Zoological Garden, were kept in large boxes, about four and a half feet high, covered on top with removable wire network, and well-ventilated through wired openings below. They were of course furnished with water, and if they declined to eat, were fed at intervals, by artificial means, with raw beef chopped fine, and passed down into the belly of the snake through a large glass-tube. Under this treatment the deaths were fewer, and the supply of venom far better. Probably this method could be usefully employed in zoological gardens, where many snakes are lost owing to their indisposition to feed during the early months of captivity.

On all occasions, for forced feeding, or for the purpose of extracting venom, the snakes were caught and held in the snake loop, Fig. 1. This is merely a staff, having a leather strap so arranged that it can be drawn out into a loop in which the serpent's neck is noosed, and so held. With this simple means all risk is avoided, and with it serpents of any size and strength to be met with among our Thanatophidiæ can be safely held and easily manipulated.

For whatever reasons the study of snake venoms had not greatly advanced since

INTRODUCTION

the last research of Fayrer and Lauder Brunton until the authors of this paper resumed the work in 1882. One of them (Dr. Mitchell) had long felt that it would be well to revise the toxicology of our American serpents which he had begun in 1858, and as the later English observers had in some points differed from

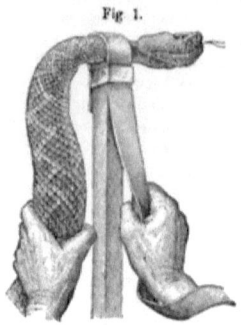

Fig. 1.

him, to learn if they or he were correct, or whether the divergence as to results was due to variations in the qualities of the venoms employed. Then too he had become conscious of certain errors in his former researches, and wished to aid in correcting them, and in filling up some of the gaps left in this branch of toxicology by himself and others.

The authors started with a theory long held by Dr. Mitchell that snake venoms are not simple in composition, but composed of two or more poisonous substances, and that in the qualities and quantities of these agents would be found an explanation of the differences between serpent venoms as to power to kill and mode of causing death.

How fertile has been the germinal idea of this research must be judged of by this present essay; which will, we trust, by lending thought and experiment in new directions hasten the day when we shall be able to treat with success the wretched thousands who now perish annually by snake-bite in India and elsewhere.

Some of our earlier results were so soon talked of and even noted in public prints, that it seemed wise for this, and all other reasons, to state what we then knew. This was done in a "Preliminary Report to the United States National Academy of Sciences, in April, 1883." In this brief essay we announced our proofs of the complex nature of snake poisons. The report was incomplete, and in the light of our present more elaborate essay may be seen to contain several erroneous statements.

It is not in the nature of things, that a research along such varied lines as our present volume follows, though extending over several years, should be perfect in detail, or complete for all genera of Thanatophidians. It is our earnest

INTRODUCTION.

hope that it will be complemented and supplemented by some of the able staff of the British Army Medical Service in the East Indies. There, only, is it possible to find enough serpents, and all the various species which it will be desirable to review toxicologically from the new stand-point which we think we have established.

We have forborne to overload this paper with comments on the later researches of others, and have made the discussion of our own work as brief as was consistent with clearness.

In writing of the various substances contained in venoms, we have given them names which are fairly descriptive, but which, as in the case of the peculiar peptone of Cobra, may perhaps excite criticism. Yet, however unsatisfactory our method of nomenclature may be, any other plan of naming the curious bodies in question would certainly have been even more misleading.

CHAPTER I.

PHYSICAL CHARACTERISTICS OF VENOM.

Physical Characteristics of Venom.—All serpent venoms are more or less alike in appearance when fresh. They are fluids varying in color from the palest amber tint to a deep yellow. Dr. Wall describes the Cobra venom as being occasionally colorless. This peculiarity we have never seen in the fresh poison of any of our serpents, except once in the coral snake; nor can the venom of one kind of snake be distinguished with certainty by any physical peculiarity from that of any other, however remote they may be in the scale of being.

When a fluid venom is allowed to dry slowly it presents no specific distinctive appearances. If desiccated too rapidly, it may look a little more gray and opaque than is common, but usually it dries into a beautifully cracked mass, deceptively like an aggregation of crystals, and which is well represented in Fig. 2.

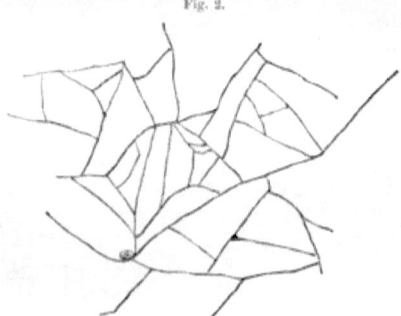

Fig. 2.

In this state it is in solid yellow particles, very fragile, bright yellow, transparent or translucent, and seemingly indestructible by time, since the dried venom of the rattlesnake, for twenty-two years in Dr. Mitchell's possession, proved as poisonous as that removed yesterday. It is equally unaltered by solution in glycerin, which keeps it permanently in unchanged toxic force, as we shall hereafter point out.[1] Neither does it appear to be injured when dry by mingling it with pure alcohol. In fact any of these three means, desiccation, glycerin, or alcohol, preserves it well.

When fresh venom of any serpent is examined with the microscope it often presents a variety of floating bodies which seem to be much alike in all cases, and are very well shown in the plates of Dr. Mitchell's former paper and in Vincent Richards's reports. In healthy serpents, but lately caged, there are fewest of these solid ingredients, as has been noticed by Richards, by Wall, and by S. Weir Mitchell. The question of the toxicity of these suspended solids has again drawn our attention to them, and we have had yet more careful and repeated microscopic examinations made by Prof. Formad. He found, like other observers, that the venom of the more vigorous snakes has the least visible solid matter; but, as in the use of the fang, the mucus and floating solids of the mouth must be considered, and, as in collecting venom from the snake, more or less of the mouth fluids mingle with the venom, it was thought well to reconsider the nature of the floating solids from the point of view of toxic activity. For the better study of the solids found in venoms we examined numerous specimens, and placed many of these in the hands of Prof. Formad, from whose notes we select the following observations:—

A drop of fresh venom, taken directly from the *Crotalus adamanteus*, was examined with a $\frac{1}{12}$ Zeiss. homog. immersion lens; amplification 800 diameters. The most striking appearance which first meets the eye is a granular material scattered about in masses of various size and shapes, resembling those formed by bacteria. There are also seen, in some cases, a few oval nucleated red blood-corpuscles, some leucocytes resembling salivary corpuscles, and others corresponding to ordinary white blood-corpuscles, the latter cells in an active state of amœboid motion. There were also observed several club-shaped epithelial cells covered with fine granular material.

The granular matter first mentioned, and which seems to form the main solid constituent of the venom, consists of two elements: Larger granules of an animal or albuminous character, and a fine granular material of vegetable nature. The albuminoid material is made up of minute particles ovoid, or somewhat irregularly angular in shape, measuring about $\frac{1}{25000}$ of an inch in their longest diameters. These ovoid particles are grouped side by side, from two to twenty in each collection, and are arranged so as to form single or double rows, or more often aggregated into irregularly shaped clusters, which vary in size from $\frac{1}{250}$ to $\frac{1}{5000}$ of an inch; the smaller masses predominating. The particles just described are colorless, refracting, and in general give the impression of bacteria. They are, however, distinguished from the latter in that they do not multiply in cultures, or respond to the aniline dye test for bacteria.

There are usually numerous bacteria in perfectly fresh venom. All the smaller particles and granular material are micrococci, measuring on an average $\frac{1}{25000}$ of an inch in diameter, are perfectly round or somewhat ovoid, and occurring singly,

---

[1] Dr. Mitchell possessed a glycerin solution which was toxic after twenty years.

in pairs, or in zoogloea masses. They are less refracting, and paler than the albuminoid particles described above, and respond promptly to the usual tests for bacteria, viz: They multiply rapidly and absorb well the aniline dyes, thus forming a marked contrast side by side with the animal granular material, which was readily discolored under the influence of acid.

The epithelial cells seen in the venom are, as a rule, few in number, are squamous or club-shaped, and in size not exceeding that of the red blood-corpuscle of the serpent. Leucocytes are also few in number, and, as well as the epithelium, are mostly covered with micrococci. A few of the white blood-corpuscles do not appear to contain micrococci, and in fresh venom, especially upon the warming stage, exhibit a quite active amœboid motion. The venom of the moccasin presents the same appearances.

If fresh venom stands but a short time exposed to the air the micrococci multiply with remarkable rapidity, forming large, pale, motionless clouds; but, in addition, multitudes of movable bacteria (the *Bacterium termo* and a bacillus— probably *Bacillus subtilis*) gradually make their appearance.[1]

The globulous masses, above described, may be collected by filtration, but as this is often a difficult or even an impossible process with a fluid as viscous as pure venom, and, as much is lost in the filter, another method was devised, and thereafter frequently used by us as an assistance in venom analysis. A tube, about 5 millimetres wide and 200 to 400 m. m. long, has a bulb blown on it midway, or at the top, and is then closed above in the blowpipe flame, and strongly heated throughout. While hot, the lower end drawn to a point, is in like manner sealed. After being cooled the tip is broken within fresh venom, which is forced up into the tube by atmospheric pressure. The end of the tube is then once more adroitly sealed in the flame.

Thus prepared the tube is suspended, so that the solids of all forms settle in a few days, while for this time, at least, the venom undergoes no such putrefactive change as is inevitable when it is exposed to the air at our ordinary spring or summer temperatures.

The solids, thus collected below, are easily separable from the supernatant venom by breaking off the two ends of the tube and allowing the precipitate to escape, with a minimum amount of liquid, from which washing in water easily separates them.

The physical appearances of the venoms of the moccasin or of the rattlesnake, thus secluded from the air in these partial vacuum tubes, undergo some curious changes of much interest.

The yellow coloring matter disappears from below upwards, and at last is seen only at the top, where the venom is in contact with the small amount of air left in the tube. At first, this change was presumed to be simply the rising of a pigment of lesser gravity. But it was noticed that the layer of yellow was of no deeper tint in its lessened bulk than when diffused. The fluid below it was left as

---

[1] Fresh venom, putrefied from long standing, appears to lose at least a portion of its virulence. But this is a point which is open to further observation.

clear and tintless as water; but when re-exposed to the air once more became yellow throughout, within one or two hours.

The yellow pigment of Cobra poison, when the dry poison was dissolved in water, does not rise in the tube or disappear, but remains unaltered. It is desirable to repeat these observations with fresh Cobra venom.

The cause of the disappearance and reappearance of the coloring matter of venom we have not been able to explain to our satisfaction, and it is one of the questions left open for inquiry.

The Specific Gravities of Venoms.—The specific gravity of the venoms of our own serpents is as follows:—

Crotalus horridus	1.054
Crotalus atrox	1.077
Crotalus adamanteus	1.061
Ancistrodon piscivorus	1.032

The specific gravity of Cobra venom is given by Wall at 1.058.
As to that of the Indian viper we can find no statement.
The losses of venom on drying were as follows:—

C. adam.	25.15	per cent.
C. atrox	25.16	"
Ancis. piscivorus	27.42	"

CHAPTER II.

THE CHEMISTRY OF VENOMS.

THE presence of alkaloids in venom, and especially of the ptomaines, has been suspected, and these bodies have been repeatedly sought for in vain. Gautier is the only chemist we recall who asserts that he found a ptomaine in a venom (Cobra). He does not state his processes, and we have been utterly unable to substantiate his statements. Lest we should in some way have erred in the conduct of this part of our labor, we asked Prof. Wolcott Gibbs to examine Crotalus venom with a view to detection of such a body. As regards this search he makes the following statement:—

"My investigation of rattlesnake venom had for its special object the comparison of the venom with the higher alkaloids. As the quantity of material at my command was small, I was obliged to content myself with the application of the ordinary tests used for the detection of alkaloids, as, for example, phospho-tungstates and phospho-molybdates, iodide of mercury and potassium, etc. etc. In many cases precipitates were obtained, but these were in no case distinctly crystalline. They resembled, on the contrary, the precipitate formed by sodic phospho-tungstate in solutions of albuminates in acetic acid. It seems, therefore, very improbable that the venom contains an alkaloid in the sense in which that term is commonly employed by chemists. On the other hand, it may still be *basic* in character, even if it be classed with albuminoids, since these are known to combine with platinous cyanide and with salicylic and other acids, exhibiting the properties of weak bases as well as of weak acids."

Venoms are of acid reaction, but when neutralized we have not observed any precipitate in specimens of these poisons.

When venom is taken from the Crotalus or Ancistrodon there is often observed in the clear poison some insoluble whitish, granular matter, which soon settles to the bottom.

The Insoluble Precipitate.—This insoluble matter, which we term the insoluble precipitate, can be collected for examination by allowing the venom to stand in hermetically sealed vertical tubes, as previously described. The precipitate soon settles to the bottom, the clear venom is then carefully drawn off, and the precipitate is repeatedly washed with distilled water and collected; the washing process is repeated until there is no trace of proteid reaction in the wash-water, or, in other words, until all of the soluble portion of the venom has been completely washed from the precipitate.

When examined under the microscope this precipitate consists of irregular masses of granular matter with epithelial cells and salivary corpuscles, and a few flat crystals resembling cholesterin.

The precipitate gives no proteid reactions with the usual proteid color tests, is insoluble in neutral saline solutions, and in weak or strong acids or alkalies. Boiling seems to render the mixture clearer.

When injected into pigeons this precipitate does not appear to possess any toxic properties.

The Globulins.—If, after the separation of the above insoluble precipitate, the venom be mixed with water and placed in a dialyser over running water it will be found that within a few hours a whitish precipitate will occur within the dialyser, and should dialysis be continued sufficiently long the precipitate will have become deposited in abundance. If the precipitate thus formed be collected on a filter it will be found that all of the coagulable proteids have been thrown down, since the filtrate now yields no coagulum by brief boiling, although it gives a proteid reaction.

The *precipitate* is now washed from the filter and subjected to repeated washings and decantations with distilled water, until the wash-water gives no proteid reaction. This purified precipitate is found to give reactions peculiar to the *globulins*; it is insoluble in distilled water, soluble in dilute neutral saline solutions, soluble in dilute acids and alkalies, becomes turbid at about $60°$ C., and is fully coagulated at a point a little above $70°$ C.

The *filtrate* still contains some proteid in solution, since we find, by the usual color and chemical tests, a proteid reaction, although it is observed that no coagulation occurs by momentary boiling. The filtrate is not precipitated by strong or weak mineral acids, by solutions of **ferric chloride or cupric sulphate**, it is precipitated but not coagulated by absolute alcohol, and if placed in a dialyser it will be found to be readily dialysable. These reactions it will be observed place the proteid which remains in solution in the filtrate among the *peptones*. But we shall revert to this hereafter.

It will thus be clear that we have separated in venom representatives of two distinct classes of proteids, one of which is insoluble in distilled water and coagulated in solution by boiling, and another which is soluble in distilled water and non-coagulable by brief boiling; the former belonging to the *globulins* and the other to the *peptones*.

The substance, however, which we find belonging to the globulins is a complex body in its composition, since, by appropriate processes, it can be resolved into three distinct principles, each of which is a globulin, but each having some properties different from its fellows. In order to distinguish these principles we have named them *water-venom-globulin*, *copper-venom-globulin*, and *dialysis-venom-globulin*, the names indicating the principal feature of the processes by which they are isolated from each other. As there are some differences in the reactions of similar principles in different species of venoms, we shall at first speak only of the venom of the *Crotalus adamanteus*.

Water-venom-globulin.—We have already stated that when a solution of the fresh or dried venom in distilled water is allowed to stand for some time, especially if the quantity of water be comparatively large, a whitish precipitate occurs which settles to the bottom of the glass, leaving in the course of a few hours a perfectly clear supernatant liquid. If sufficient water has been added at first, the addition of more distilled water to the supernatant liquid will not cause any further precipitate.

The precipitate is now collected and repeatedly washed with distilled water and decanted until the wash-water yields no proteid reaction.

The following gives the results of some of the many reactions upon the addition of the various reagents used:—[1]

Decided reactions with the usual proteid tests.
Boiling—causes coagulation.
Sodic chloride (0.75 per cent.)—slightly soluble.
   '10 ")—soluble, forming a turbid solution; the solution is not precipitated by carbonic acid[2] nor by the addition of ether.
    —boiling the solution causes coagulation.
    —the solution is precipitated by saturation with sodic chloride.
Carbonic acid[2]—soluble.
Sodic carbonate—very soluble; solution not precipitated by carbonic acid.
Hydrochloric acid (0.4 per cent.)—very soluble.
Metaphosphoric acid—insoluble.
Orthophosphoric acid—dissolves
Sodic metaphosphate—insoluble.
Sodic orthophosphate—very soluble.
Potassic sulphate—very soluble.
Calcic chloride—very soluble.
Acetic acid (5 per cent.)—very soluble.
Acetic acid (glacial)—very soluble.
Coagulation occurs at about 64–73° C.

Since this body is precipitated by saturation with sodic chloride, and dissolves with difficulty in a 0.75 per cent. solution of sodic chloride, it seems more akin to myosin than other of the globulins.

The Copper-venom-globulin.—After the separation of the water-venom-globulin the filtrate gives well-marked proteid reactions and decided coagulation by boiling. If now a few drops of cupric sulphate (10 per cent.) be cautiously added a second precipitate will occur, and which can be separated as in the previous instance. In adding the cupric sulphate great caution must be exercised lest too much be added with the result of a complete or partial re-solution of the precipitate.

The precipitate is sometimes comparatively slight at first, increasing upon standing, and complete within about twenty-four hours. The clear filtrate should give no precipitate after the addition of a small amount of the copper solution and after standing twenty-four hours longer.

---

[1] In all of these reactions with the globulins, unless otherwise apparent, about 1 c. c. of the suspended globulin in distilled water was placed in a small test-tube, and from one to two drops of standard laboratory solutions of reagents were allowed to run down the inside of the tube.

We have made a large number of tests with various reagents, and from this number have selected only such as will serve us some purpose in distinguishing these different bodies.

[2] Where carbonic acid is used in these tests we have reference to the super-saturated carbonic acid water (soda water) of commerce.

The precipitate thus obtained is washed as in the preparation of the water-venom-globulin, and when thus purified it does not give any color reaction with the ammonia or the ferrocyanide and acetic-acid tests for copper, and therefore cannot be regarded as a salt of this metal.

The *copper-venom-globulin* gives the following reactions:—

Decided reactions with the usual proteid tests.
Sodic chloride (0.75 per cent.)—insoluble.
            (10 ")—insoluble.
                  —the addition of crystals of sodic chloride seems to dissolve it slightly; this solution is cleared somewhat by boiling; the same effect by boiling the suspended mixture; the clearing is no doubt the result of the formation of coagula.
Carbonic acid—insoluble.
Sodic carbonate—very soluble, forming a beautiful clear solution; boiling has no effect; the solution is precipitated by carbonic acid.
Hydrochloric acid (0.4 per cent.)—exceedingly soluble.
Metaphosphoric acid—insoluble; boiling no effect.
Orthophosphoric acid—very soluble, forming an absolutely clear solution; boiling has no decided effect.
Sodic metaphosphate—insoluble; boiling no effect.
Sodic orthophosphate—soluble in a much larger amount than is necessary in dissolving the water-venom-globulin; boiling has no effect, unless to clear the solution some.
Potassic sulphate—insoluble; boiling no effect.
Calcic chloride—less soluble than water-venom-globulin.
Acetic acid (5 per cent.)—very soluble.
Acetic acid (glacial)—very soluble.

The *Dialysis-venom-globulin*.—The filtrate, after the separation of the water-venom-globulin and copper-venom-globulin, still gives a decided amount of coagula by boiling, and also all of the characteristic color reactions for proteids. If the filtrate be now subjected to dialysis, best by means of a large dialyser placed over running water, in the course of twenty-four hours a considerable amount of precipitate will be deposited within the dialyser, and which may be collected on a filter, and repeatedly washed as in the preparation of the preceding globulins.

If dialysis is carried on for a sufficient length of time the whole of this principle will be precipitated, since the filtrate from the globulin will give no coagula by boiling, nor any precipitate by strong nitric acid. A proteid still remains in solution, however, which has been already alluded to as being a peptone. This body being less dialysable than the salts which hold the globulins in solution, still remains in part within the dialyser, even when the salts are so fully withdrawn as to entirely precipitate the globulins.

The *dialysis-venom-globulin* gives the following reactions:—

Decided reactions with the usual proteid tests.
Sodic chloride (0.75 per cent.)—insoluble.
            (10 ")—slightly soluble.
                (crystals)—more soluble, forming a very cloudy solution; boiling clears the solution some; the same degree of clearing does not occur in the mixture without the sodic chloride.
                —the addition of carbonic acid to the solution with crystals causes a beautiful clear solution, which is made cloudy by boiling.

THE CHEMISTRY OF VENOMS. 13

Carbonic acid—soluble; cloudiness by boiling.
Sodic carbonate—very soluble; boiling no effect.
Hydrochloric acid (0.4 per cent.)—very soluble.
Metaphosphoric acid—rendered of a yellowish tint; not appreciably dissolved; boiling no appreciable effect.
Orthophosphoric acid—very soluble; boiling no effect.
Sodic metaphosphate—very soluble, forming a very clear solution; boiling no effect.
Sodic orthophosphate—slightly soluble; dissolving slowly in excess, forming a slightly turbid solution; boiling clears absolutely.
Potassic sulphate—insoluble; boiling no decided effect.
Calcic chloride—soluble by the addition of a comparatively larger amount; boiling causes coagulation.
Acetic acid (5 per cent.)—very soluble.
Acetic acid (glacial)—very soluble.

The Venom Peptone.—After the separation of the dialysis-globulin the filtrate, as before stated, gives no coagula by brief boiling, but by testing with the usual proteid tests very decided reactions are obtained. It is further found that if the above filtrate is placed in a fresh dialyser, that the principle giving the proteid reactions will readily pass through the membrane. The fact that this substance will dialyse readily, and that it is not immediately coagulated at the temperature of boiling water, and not precipitated by cupric sulphate and ferric chloride, nor by neutralization, renders it certain that it belongs to a peculiar class of bodies which are known as peptones, and which are ordinarily the result of peptic or tryptic digestion. This peptone may also be prepared by briefly boiling the solution of venom, which coagulates the other albuminous principles, and leaves this in solution; but the coagula caused by boiling the solution of Crotalus are so extremely fine, that it is impossible to filter the mixture clear, even by repeated filtration through many thicknesses (7) of the best filter paper; furthermore, continued boiling causes a breaking down of the peptone with the apparent formation of fine coagula (see Cobra peptone, p. 17). We, however, prepared the peptone by dialysis, and obtained the following reactions:—

No immediate coagulation at a temperature of 100° C.
Full reactions with the proteid color tests.
No precipitate with weak or strong nitric acid.
Ferric chloride—no precipitate.
Cupric sulphate—no precipitate.
Mercuric chloride—decided precipitate.
Absolute alcohol—precipitate; precipitate redissolved by the addition of water.
Mercuric nitrate—decided precipitate.
Potassic hydrate—precipitate by saturation; precipitate redissolved by the addition of nitric acid, forming a decidedly yellowish solution, which becomes decolorized by further addition of acid.
Potassic ferrocyanide in presence of weak acetic acid—a precipitate.

To revert now to the globulins and their distinctive features, it seems clear that these principles must exist in the venom as distinct bodies, and are not simply representatives of a single globulin which have arisen through our manipulations. The first distinguishing feature between them is represented in the process of isolation, but if we place the reactions of the different globulins in parallel columns, we find that, while they have very close resemblances, as they naturally should since they are so intimately related, they are very readily distinguished from each other. The properties of all globulins are so readily affected by even the simplest manipulations that it is likely that mere precipitation may affect them in regard to their solubility, while drying may completely destroy this property. Having these facts in mind, it seems almost a necessity that the processes through which we put these globulins, in order to get them isolated in a pure state, has more or less modified their chemical, and possibly their physiological properties.

The tests made with these globulins were all made at different times, the one globulin was examined one day, and another on another day, so that the reactions given are not absolutely accurate as a matter of comparison, but only relative, since the standard of solubility, which was of course an arbitrary one, was simply carried in the mind throughout the examinations. We believe, however, that they are practically correct.

Reagent.	Water-venom-globulin.	Copper-venom-globulin.	Dialysis-venom-globulin.
Sodic chloride (10 p. c.)	Soluble	Insoluble	Slightly soluble,
Carbonic acid	Soluble	Insoluble	Soluble.
Sodic carbonate	(Very soluble; not precipitated by CO_2	Very soluble; precipitated by CO_2)	Very soluble.
Hydrochloric acid (0.4 p. c.)	Very soluble	Very soluble	Very soluble.
Metaphosphoric acid	Insoluble	Insoluble	(Insoluble; rendered of a yellowish tint.
Orthophosphoric acid	Soluble	Very soluble	Very soluble.
Sodic metaphosphate	Insoluble	Insoluble	Very soluble.
Sodic orthophosphate	Very soluble	Less soluble	Still less soluble.
Potassic sulphate	Very soluble	Insoluble	Insoluble.
Calcic chloride	Very soluble	Less soluble	Less soluble.
Acetic acid (5 per cent.)	Very soluble	Soluble	Very soluble.
Acetic acid (glacial)	Very soluble	Soluble	Very soluble.

The venom of the Moccasin (*Ancistrodon piscivorus*) was subjected to an analysis similar to that of the Crotalus, the isolated proteids giving the following reactions:—

Water-venom-globulin.

Decided reactions with the usual proteid color tests.
Boiling—clears the mixture without the apparent formation of any coagula.
Sodic chloride (0.75 per cent.)—insoluble.
   (10 ")—somewhat soluble, solution not absolutely clear; boiling clears absolutely without the formation of coagula.
   (crystals)—somewhat soluble; solution not precipitated by carbonic acid.
Carbonic acid—insoluble.
Sodic carbonate—soluble, forming slightly turbid solution; boiling clears the solution without giving coagula; the addition of crystals of sodic chloride to the hot boiled solution causes a precipitate, this precipitate being coagulated by boiling.
Hydrochloric acid (0.4 per cent.)—somewhat soluble.
   (5 ")—soluble.
Metaphosphoric acid—insoluble.
Orthophosphoric acid—soluble.
Sodic metaphosphate—slightly soluble; solution rendered clearer by boiling.
Sodic orthophosphate—soluble; solution rendered absolutely clear by boiling.

THE CHEMISTRY OF VENOMS.

Potassic sulphate—soluble; solution rendered absolutely clear by boiling.
Calcic chloride—soluble; solution rendered clearer by boiling.
Acetic acid (5 per cent.)—insoluble.
Acetic acid (glacial)—insoluble?

Copper-venom-globulin.

Boiling—clears somewhat; no coagula.
Sodic chloride (0.75 per cent.)—insoluble.
      (10 ")—insoluble.
      (*crystals*)—insoluble; boiling partially clears without the formation of any coagula.
Carbonic acid—somewhat soluble; boiling clears absolutely.
Sodic carbonate—very soluble; boiling no effect.
Hydrochloric acid (0.4 per cent.)—very soluble.
Metaphosphoric acid—insoluble; boiling appears to clear slightly.
Orthophosphoric acid—very soluble.
Sodic metaphosphate—insoluble; boiling clears somewhat.
Sodic orthophosphate—somewhat soluble; boiling clears beautifully.
Potassic sulphate—insoluble; boiling clears slightly.
Calcic chloride—slowly dissolved; not so soluble as water-globulin; boiling gives a slight cloudiness.
Acetic acid (5 per cent.)—soluble.
Acetic acid (glacial)—soluble.

Dialysis-venom-globulin.

Boiling—clears almost absolutely without the apparent formation of coagula; boiled solution precipitated by saturation with crystals of sodic chloride.
Sodic chloride (0.75 per cent.)—insoluble.
      (10 ")—somewhat soluble; dissolves slowly, forming a slightly turbid solution; boiling seems to clear some without the formation of coagula.
Carbonic acid—very soluble; slight turbidity by boiling.
Sodic carbonate—very soluble; boiling no effect.
Hydrochloric acid (0.4 per cent.)—very soluble.
Metaphosphoric acid—slightly soluble; yellowish tint; boiling clears slightly with the formation of coagula.
Orthophosphoric acid—very soluble; boiling no effect.
Sodic metaphosphate—insoluble.
Sodic orthophosphate—soluble; boiling no effect.
Potassic sulphate—somewhat soluble.
Calcic chloride—very soluble, form a beautiful clear solution; boiling causes slight turbidity.
Acetic acid (5 per cent.)—soluble.
Acetic acid (glacial)—soluble.

Moccasin Peptone.

1. Readily dialyzable.
2. Not immediately coagulated at a temperature of 100° C., but gradually coagulated by prolonged boiling (see Cobra peptone, p. 17).
3. Reaction with the xantho-proteic test (nitric acid and **ammonia**)
4. Reaction with Millon's reagent (mercuric nitrate)
5. No precipitate with weak or strong nitric acid.
6. No precipitate with CO_2.
7. No precipitate with ferric chloride.
8. No precipitate with cupric sulphate.

9. Precipitated by mercuric chloride.
10. Precipitated by absolute alcohol.
11. Gives a faint reddish tinge with a strong solution of potassium hydrate, and a trace of cupric sulphate.
12. Not precipitated by strong acetic acid (glacial).
13. Precipitated by very dilute acetic acid; precipitate being redissolved by further addition of acid.
14. Full reaction with Adamkiewicz's test for proteids.
15. Precipitated by adding a large quantity of sodium chloride, the precipitate being redissolved on the addition of a large quantity of glacial acetic acid.
16. Precipitated by mercuric nitrate.
17. Precipitated by absolute alcohol; precipitate being apparently redissolved on the addition of water.
18. Precipitated by saturation with potassium hydrate; precipitate being redissolved by the addition of nitric acid, with the formation of a decidedly yellow solution (xantho-proteic) which becomes decolorized by addition of acid.
19. Precipitated by potassium ferrocyanide in the presence of weak acetic acid.

Venom-peptone by dialysis gives identical reactions.

For convenience of comparison we append here in parallel columns the principal reactions of the Moccasin globulins, remembering in this connection the difference in the properties manifest in their methods of preparation.

Reagent.	Water-venom-globulin.	Copper-venom-globulin.	Dialysis-venom-globulin.
Boiling	Clears almost absolutely	Clears some	Clears some.
Sodic chloride (10 per cent.)	Somewhat soluble	Insoluble	Somewhat soluble.
Carbonic acid	Insoluble	Somewhat soluble	Very soluble.
Sodic carbonate	Soluble	Very soluble	Very soluble
Hydrochloric acid (0.4 p. c.)	Somewhat soluble	Very soluble	Very soluble.
Metaphosphoric acid	Insoluble	Insoluble	Slightly soluble.
Orthophosphoric acid	Soluble	Very soluble	Very soluble.
Sodic metaphosphate	Somewhat soluble	Insoluble	Insoluble.
Sodic orthophosphate	Soluble	Less soluble	Soluble.
Potassic sulphate	Soluble	Insoluble	Slightly soluble
Calcic chloride	Soluble	Insoluble	Very soluble.
Acetic acid (5 per cent.)	Insoluble	Soluble	Soluble.
Acetic acid (strong)	Insoluble	Soluble	Soluble.

For reactions of the peptones of the various venoms see p. 19

Cobra Venom.—We have been able to isolate in Cobra venom only two proteids, and these correspond in their characters to the two types of proteids found in the venoms of the Crotalus and Ancistrodon. In other words, we have isolated a *globulin* and a *peptone-like* principle. The globulin we are able to precipitate completely by the addition of a proper amount of distilled water, after which the solution gives no coagulum by boiling. There is then left in solution a proteid, which evidently belongs to the peptones, although giving some extraordinary reactions.

The venom-globulin thus isolated and purified, as in the preparation of the globulins previously mentioned, possesses the peculiar properties of the globulin family, and, in accordance with our nomenclature, since it is entirely precipitated by the addition of distilled water, is a *water-venom-globulin.*

The following are some of the reactions given by this substance (the *water-venom-globulin* suspended in distilled water):—

Boiling—coagulates.
Sodic chloride (0.75 per cent.)—insoluble.
        (10 ")—soluble; boiling gives slight turbidity.
                      —sodic chloride solution apparently unaffected by carbonic acid.
Carbonic acid—insoluble.
Sodic carbonate—soluble, slightly turbid solution; boiling makes perfectly clear.
Hydrochloric acid (0.4 per cent.)—soluble.
Metaphosphoric acid—insoluble; boiling no appreciable effect.
Orthophosphoric acid—very soluble; boiling makes solution absolutely clear.
Sodic metaphosphate—insoluble; boiling no appreciable effect.
Sodic orthophosphate—somewhat soluble; boiling renders perfectly clear.
Potassic sulphate—somewhat soluble.
Calcic chloride—soluble; opalescence of solution increased by boiling.
Acetic acid (5 per cent.)—soluble.
Acetic acid (glacial)—soluble.

Cobra-venom-peptone.—The venom-peptone from Cobra may be prepared by boiling, thus coagulating the globulin, or by dialysis. Great difficulty is experienced in the former process, since the coagula are so fine that it is impossible, save in rare instances, to obtain a clear filtrate, and as to these we have no explanation to offer for the exception. The peptone prepared by boiling or by dialysis gives identical reactions.

Before detailing the reactions of this body it may be well to notice a peculiar property exhibited by all venom-peptones which gives them a very distinguishing feature. After boiling the venom for a few minutes and then filtering, the filtrate will again give further coagula by continued boiling, and so the process of boiling and filtering, and reboiling the filtrate may go on repeatedly, yet the clear filtrate will in every instance give fresh coagula. Indeed the boiling process may be continued for an hour or more, and yet at the end of that time the filtrate will still yield coagula. However, after the venom solution has once been boiled, coagulation does not recommence in the filtrate until it has been boiled for a few moments. These most interesting facts suggest that the coagula formed after the first boiling are due to a gradual decomposition of what is in some sense a non-coagulable proteid, since coagulable proteids all coagulate at once and completely when a definite temperature is reached; the coagula which follow repeated or prolonged boiling appear to be due to such a decomposition of proteids as violent chemical or physical action could alone account for.

It seems to us perfectly clear that the body which is thus gradually broken up by prolonged boiling is a *peptone*. Our principal reasons for this belief are that the body so coagulated is very readily dialysable, is not precipitated by ferric chloride, or cupric sulphate, and in the case of the Cobra is not precipitated by absolute alcohol, or *mercuric chloride*, is not coagulated below the boiling point, and in fact not until boiling has gone on for a few moments. The following reactions seem to be sufficiently characteristic.

These results we obtained from a solution of the *Cobra-venom-peptone* obtained
3   April, 1886.

by dialysing venom for forty-eight hours. The dialysate was perfectly clear and neutral in reaction:—

Boiling—no result until after a few moments, when it becomes cloudy, the cloudiness increasing as boiling continues; strong nitric acid dissolves the precipitate.
Color reactions for proteids—the xantho-proteic, Millon and Biuret reactions are all obtained.
Ferric chloride—no effect.
Cupric sulphate—no effect.
Mercuric chloride—no effect.
Mercuric nitrate—decided precipitate.
Absolute alcohol—no precipitate.
Potassic ferrocyanide + weak acetic acid—precipitate.
Nitric acid (strong)—no precipitate.
Hydrochloric acid (strong)—no precipitate.
Acetic acid (strong)—no precipitate.
Sodic chloride (saturation)—precipitate; acetic acid, large quantity, dissolves.
Potassic hydrate to saturation—precipitate.
Tannic acid—decided precipitate.
Basic acetate of lead—decided precipitate.

Several very remarkable facts are the coagulation by prolonged boiling and the non-precipitation by mercuric chloride and absolute alcohol. Since this peptone is precipitated by weak acetic acid in the presence of potassic ferrocyanide it has a slight resemblance to Meissner's A peptone, although materially differing, as some of the above reactions show, from any other described body of this class.

As a matter of some interest, it is desirable to know if similar globulins in different venoms are identical in their chemical nature, or whether they give any reactions which may distinguish them. We have accordingly, as in previous cases, placed the reactions of the corresponding globulins side by side.

I. *Water-venom-globulin.*

Reagent.	Crotalus horridus.	Ancistrodon piscivorus.	Cobra.
Boiling	Coagulates	Apparently dissolves	Coagulates.
Sodic chloride (10 per cent.)	Soluble	Somewhat soluble	Soluble.
Carbonic acid	Soluble	Insoluble	Insoluble.
Sodic carbonate	Soluble	Soluble	Soluble.
Hydrochloric acid (0.4 p. c.)	Soluble	Somewhat soluble	Soluble.
Metaphosphoric acid	Insoluble	Insoluble	Insoluble.
Orthophosphoric acid	Soluble	Soluble	Soluble.
Sodic metaphosphate	Insoluble	Somewhat soluble	Insoluble.
Sodic orthophosphate	Very soluble	Soluble	Somewhat soluble.
Potassic sulphate	Very soluble	Soluble	Somewhat soluble.
Calcic chloride	Very soluble	Soluble	Soluble.
Acetic acid (5 per cent.)	Soluble	Insoluble	Soluble.
Acetic acid (strong)	Soluble	Insoluble	Soluble.

II. *Copper-venom-globulin.*

Reagent.	Crotalus horridus.	Ancistrodon piscivorus.
Boiling	Coagulates	Apparently dissolves.
Sodic chloride (10 per cent.)	Insoluble	Insoluble.
Carbonic acid	Insoluble	Somewhat soluble.
Sodic carbonate	Very soluble	Very soluble.
Hydrochloric acid (0.4 p. c.)	Very soluble	Very soluble.
Metaphosphoric acid	Insoluble	Insoluble.
Orthophosphoric acid	Very soluble	Very soluble.
Sodic metaphosphate	Insoluble	Insoluble.
Sodic orthophosphate	Soluble	Soluble.
Potassic sulphate	Insoluble	Insoluble.
Calcic chloride	Soluble	Insoluble.
Acetic acid (5 per cent.)	Soluble	Soluble.
Acetic acid (glacial)	Soluble	Soluble.

III. *Dialysis-venom-globulin.*

Reagent.	Crotalus adamanteus.	Ancistrodon piscivorus.
Boiling	Coagulation	No coagulation ?
Sodic chloride (10 per cent.)	Somewhat soluble	Somewhat soluble.
Carbonic acid	Soluble	Very soluble.
Sodic carbonate	Very soluble	Very soluble.
Hydrochloric acid (0.4 p. c)	Very soluble	Very soluble.
Metaphosphoric acid	Insoluble	Slightly soluble.
Orthophosphoric acid	Very soluble	Very soluble.
Sodic metaphosphate	Very soluble	Insoluble.
Sodic orthophosphate	Soluble	Soluble.
Potassic sulphate	Insoluble	Slightly soluble.
Calcic chloride	Soluble	Very soluble.
Acetic acid (5 per cent.)	Soluble	Soluble.
Acetic acid (glacial)	Soluble	Soluble.

It will be noticed by a careful comparison that the corresponding principles in different venoms differ quite as much from each other as the globulins in any one variety of venom.

Venom Peptones.—We have not been able to detect any chemical differences in the venom peptones of the Crotalus and Ancistrodon. Cobra venom peptone is distinguished from that of the Crotalus and Ancistrodon by its non-precipitability by mercuric chloride and absolute alcohol.

Daboia Venom.—We have had a small quantity (a few grains) of Daboia venom at our disposal, but too little to attempt any detailed chemical investigations. In two examinations, however, with very small quantities, we separated two bodies corresponding to those in Cobra, that is a *water-venom-globulin* and a *peptone*. The former exists in exceedingly small quantity while the latter dialyses with apparently much more difficulty than that of the Cobra.

The Proportions of Proteid Constituents in Different Venoms.—An examination of good specimens of the dried venoms of the *Crotalus adamanteus*, *Ancistrodon piscivorus*, and *Cobra* gives us the following proportions of the globulins and peptones:—

Crotalus adamanteus—
0.5 gram dried venom = water-venom-globulin   0.0495
                      copper-venom-globulin 0.0375
                      dialysis-venom-globulin 0.0360
                                             ─────────
                                             0.1230 = globulins.
                                             0.3770 = peptone[1] (estimated.)

Ancistrodon piscivorus—
0.3364 gram dried venom = water-venom-globulin   0.0034
                          copper-venom-globulin 0.0182
                          dialysis-venom-globulin 0.0047
                                                 ─────────
                                                 0.0263 = globulins.
                                                 0.3101 = peptone[1] (estimated.)
According to this estimate there would be in 0.5 gram 0.0391 globulins.
                                                      0.4609 peptone.[1]

Cobra—
0.2 gram dried venom = water-venom-globulin 0.0035
                       peptone[1] 0.1965 (estimated).
According to this estimate there would be in 0.5 gram 0.0087 globulin.
                                                      0.4912 peptone.[1]

From these analyses it will be observed that the dried venom of the *Crotalus adamanteus* contains 24.6 per cent. of globulins, the Ancistrodon 7.8 per cent., and the Cobra 1.75 per cent. The globulins in the Crotalus venom appear to be in almost equal proportions, while in the Ancistrodon the copper-venom-globulin is about five times greater than the water-venom-globulin and about four times more than the dialysis-venom-globulin—the two latter being nearly in the same proportion—therefore constituting more than half of the entire weight of globulins.

These differences in the proportions of the various globulins in any specimen of venom and the differences in the proportions of globulins and peptones in different venoms are of immense importance in affording an explanation of the physiological peculiarities exhibited in poisoning by different species of snakes. It will be observed that the proportion of globulins in Crotalus is over three times the quantity in the Ancistrodon, and nearly fifteen times that in the Cobra.

---

[1] Including the salts, which are in very small quantity.

CHAPTER III.

EFFECTS OF VARIOUS AGENTS ON VENOM.

Effects of Various Agents on Venom.—The influence of acids, alkalies, and salts on venoms has been studied by several observers, with results which vary remarkably; so that for this and for other reasons there is still room for research of this nature. The questions thus brought up have a twofold interest, the one chemical and the other toxic. Numerous bodies precipitate or dissolve venoms; but among those which most plainly alter these poisons, only a few so change them as to lessen or destroy their poisonous efficiency. Unfortunately, that which alters the poison as such, is always equally destructive to the tissues of the body, and no agent as yet employed can be shown to have the power to enter the blood, and there affect the venom without doing harm to other albuminous substances. So far, we have learned only that amidst the agents which precipitate venom, there are some which weaken or annihilate its toxic force. They can be thrown into the fang tracks, and where they are made to mingle with the venom will destroy it as impartially as they do the innocent tissues in which it lies.

It may not be out of place to remark that we have made no direct study of agents as antidotes. Too much yet remains to be known of these poisons before we can hope to find a means of antagonizing them physiologically. Our local or chemical antidotes are sufficiently effective.

Effect of Desiccation of Venom.—Allowed to dry at ordinary temperatures, the venoms retain their poisonous activity almost unaltered. When again water is added they act as usual, except that, owing perhaps to imperfections in redissolution, they do not produce as much local effect within as short a time as do the fresh fluid venoms. Neither, it may be added, is the general toxic influence quite as rapid when venom has been once desiccated.

The Effects of Various Agents on the Toxicity of Venoms. Age.—Some fresh venom of the *Crotalus horridus* was dissolved in an equal quantity of pure glycerine and the vial corked and sealed in 1863. In November, 1882, the contents of the vial were examined. The solution was perfectly clear, and had at the bottom a small mass of what appeared to be a fungous growth. Some of the venom was now injected into various animals to test its toxicity. The following experiment attests its power:—

Experiment.—Pigeon. Injected, at 5:12 P. M., into the muscles of the thigh about six drops of the above glycerin solution.

    5:14. Animal decidedly weakened.
    5:25. There is considerable blackening of the tissues about the point of injection, the parts

being much swollen, the leg stiff, the muscles at the point of injection are paralyzed, and sensibility of the leg destroyed. The pigeon lies on its side unable to stand, is exceedingly prostrated, and breathes laboriously. Observation now ceased until 8 A.M. following morning, when the animal was found dead and in general rigor mortis, excepting the muscles at point of injection.

Autopsy.—The tissues were dark, congested, and suffused with serum for an area of one and a half inches from point of injection. The viscera of the thoracic and abdominal cavities appeared slightly congested; the heart was arrested in systole and contained dark clots; the blood everywhere was dark and clotted. Microscopically the muscular fibres did not appear to be greatly disorganized, although in some of the fibres no transverse striæ or nuclei could be discovered.

The Effects of Dry Heat. Experiment.—0.03 gram of dried (*Crotalus adamanteus*) venom was subjected in a dry oven to a gradually rising temperature to 83.5° C., and maintained at this point for half an hour. The venom, after cooling, was dissolved in 1 c.c. of distilled water.

2:57. Injected the above into the thigh of a pigeon.
4:49. Violent convulsions and death. Local effects decidedly marked.

Experiment.—Repeated the above, but subjecting the venom to a temperature of 100 C. for ten minutes.

3:43. Injected into the thigh of a pigeon.
6:00. No decided symptoms. On the following morning the animal was dead. The local effects were marked.

Experiment.—Repeated the above, but subjecting the venom to a temperature of 110° C. for thirty minutes.

4:46. Injected into the thigh of a pigeon.
5:25. Convulsions.
5:45. Died. The local effects were marked.

From these results it seems clear that heating the dry venom to a degree above boiling point does not apparently alter its poisonous activity. The delay in the occurrence of death in the second experiment suggests that the venom was altered, but in the third experiment in which the temperature was even higher, and this degree of heat maintained for a much longer time, death occurred even sooner than in the first experiment, showing that the differences must have been dependent upon conditions in the animals.

The Effects of Moist Heat. Experiment.—0.03 gram dried venom (*Crotalus adamanteus*) was dissolved in 1 c.c. distilled water, and gradually heated until a flocculent precipitate occurred.

This was injected into the thigh of a pigeon in the evening. The next morning the animal was found dead.

Experiment.—0.03 gram dried venom (*Crotalus adamanteus*) was dissolved in 1 c.c. distilled water and subjected to a gradually rising temperature to 50° C.

3:49. Injected the above into the breast muscles of a pigeon.
3:51. Very weak, pupils apparently contracted, trembling; breathing laborious.
4:00. Dead. At the point of injection the tissues were decidedly congested and purplish and suffused with blood. The blood generally was fluid, but some soft clots were found in the abdominal vessels.

EFFECTS OF VARIOUS AGENTS ON VENOM.   23

Experiment.—Subjected a similar amount of venom in solution to a rising temperature to 65° C.

 4:01. Injected into the breast muscles of a pigeon.
 4:05. Head depressed.
 4:10. Very weak, falls on the side.
 5:02. Dead. The local effect is not so marked as in the previous experiment. The injection was merely subcutaneous. The viscera did not appear congested or abnormal; the heart was arrested in systole; blood everywhere fluid and dark; no ecchymoses in the peritoneum; muscles appear darker than normal.

Experiment.—Repeated the above, but increasing temperature to 74° C.

 4:13. Injected into the breast muscles of a pigeon.
 4:19. Weak, falls on side.
 5:05. Dead. Blood clotted; local effect the same as in previous experiment.

Experiment.—Results the same as in the last experiment, excepting that the local effects were more marked. This animal lived a half hour longer than the last, which will probably account for the difference.

Experiment.—The same, but subjecting the solution to 76.5° C.

 4:04. Injected into the breast muscles of a pigeon.
 4:27. Unable to stand.
 6:00. Nearly dead.
 Following morning. Extremely feeble, too weak to stand; there is a muco-sanguinolent discharge from the bowels.
 Second day. Very feeble.
 Third day. Recovering.

Experiment.—The same, but subjecting the solution to 79.5° C.

 4:00. Injected into the breast muscles of a pigeon.
 5:50. No symptoms.
 Following morning animal well.

Experiment.—The same, but subjecting the solution to 81° C.

 4:31. Injected into the breast muscles of a pigeon.
 4:45. Apparently a little stupid.
 5:50. No further effect.
 Following morning. Animal well.
 Second morning. Animal well.

Experiment.—Boiled a similar amount of solution for two minutes.

 3:26. Injected into the breast muscles of a pigeon.
 4:30. No effect.
 Following morning. No effect.

The above very interesting series of experiments clearly shows that the effect of heat on a solution of venom is very positive, that the toxicity of venom is decidedly affected, and that the greater the increase of temperature between certain limits the greater is the destruction of the poisonous power of the venom. It will be observed in the second experiment, which is the first in which any positive temperature was observed, that the animal died in about *ten minutes* after injection; in the third experiment in about *one hour;* in the fourth and fifth experiments in about *three-fourths of an hour,* and *an hour and three-quarters* respectively; in the sixth experiment in about *two hours;* the animal was nearly dead, but finally recovered in the seventh experiment and in the subsequent ones there were no poisonous symptoms. It will thus be observed that there is a gradual impairment of the toxicity of the venom increasing with the increase of temperature, and that when we reach 76.5° C. we have almost reached the temperature at which toxicity seems to be completely destroyed. We say *seems* completely destroyed, because we have found that the solution is still toxic even when boiled, although there is not sufficient active poisonous matter left after boiling in the small amount of venom we used in this group of observations to cause decidedly poisonous effects in pigeons.

The results of boiling solutions of *Moccasin* and *Cobra* venoms are quite different from the above, as the following experiments clearly show:—

Experiment.—Dissolved 0.015 gram dried Moccasin in 1 c. c. distilled water, and gradually heated to 78° C.

    3:40. Injected into the breast of a pigeon.
    3:50. Rocking.
    4:45. Nearly gone; some local effect.
    Following morning the animal was dead. The local effect (darkening) was marked, but not comparable to that caused by the unboiled venom.

Experiment.—Boiled 0.015 gram dried Moccasin (*piscivorus*) dissolved in 1 c. c. distilled water for one minute.

    3:28. Injected the above into the breast muscles of a pigeon.
    3:35. Too weak to stand.
    4:15. Dead. There are no local effects.

Experiment.—Dissolved about 1½ minims of fresh Moccasin venom in about 1 c. c. distilled water, then boiled in a test-tube, filtered and injected one-half into the breast muscles of a pigeon at 4:30.

    4:55. Very slight local effect; darkening and swelling; the animal is weak and has respiratory disturbance.
    Injected the other half.
    5:00. Rocking; irregular breathing; somewhat stupefied.
    5:20. Eyes closed; stupefied; breathing irregular.
    Following morning. There was a large, light-colored, œdematous swelling (see Plate No. 1) within, which was a cavity about an inch in diameter, full of broken-down tissue, having a grayish muddy, gangrenous appearance, and a putrefactive odor, while the surrounding muscular tissues were normal in appearance.

It will be observed in this series of experiments with the Moccasin venom that there is also a very decided alteration in the poisonous properties of the venom. But here we find that although the amount of venom used was only one-half the quantity employed in the Crotalus series, boiling does not destroy its ability to kill. It will also be noticed here, as in the case of the Crotalus, that a sufficient degree of heat has an obvious effect on the power of the venom to produce the peculiar lesions at the point of injection.

The effect of heat upon solutions of *Cobra* venom is not so marked.

Experiment.—0.03 gram dried Cobra venom was dissolved in 1 c. c. distilled water and subjected to a temperature gradually rising to 74° C.

    4:10. Injected into the breast muscles of a pigeon.
    4:16. Unable to stand.
    4:20. Dead.

Experiment.—The same, excepting that the temperature was raised to 79.5° C.

    4:12. Injected as above.
    4:21. Unable to stand.
    4:25. Dead.

Experiment.—The same, solution being brought to boiling point in a test-tube.

    4:45. Injected into the breast muscles of a pigeon.
    5:00. Unable to stand.
    5:03. Convulsions followed by death.

Experiment.—0.015 gram dried venom dissolved in 1 c. c. distilled water and boiled in a test-tube for about two minutes.

    3:51. Injected into the breast muscles of a pigeon.
    4:15. Unable to stand.
    4:22. Dead. No local effects.

From these experiments it appears that the toxicity of venom is not impaired by brief heating as high as 79.5° C., the time of death being in these experiments about the same as with the unheated solution. In the last two experiments in which the solution was boiled, the time of death is delayed, especially so in the last experiment, but here it must be observed that but one-half the dose was used.[1]

In one experiment made on the venom of the Copperhead (*Ancistrodon contortrix*) the effect seemed to be in degree between that of the *Crotalus* and *Ancistrodon piscivorus*.

Experiment.—0.03 gram dried venom was dissolved in 1 c. c. distilled water and boiled in a test-tube for two minutes.

    5:00. Injected into the breast muscles of a pigeon.
    5:10. Unable to stand.
    5:20. Incoördination.
    6:00. Very weak.
    Following morning. Dead. There were very slight local effects; the blood was clotted in soft black clots; heart arrested in systole, auricles full of clots. The interior of the thoracic cavity had a mucky brownish appearance; the viscera did not appear congested, and there were no ecchymoses.

A similar dose of the unheated copperhead venom kills promptly with decided local effects. It will thus be apparent that boiling decidedly alters its toxic power.

The effect of boiling on the venom of the *Crotalophorus* is as decided as on that of the *Crotalus*.

Experiment.—Two drops of the fresh venom of the *Crotalophorus* was dissolved in 1 c. c. distilled water and boiled for a moment.

    4:58. Injected into the breast muscles of a pigeon.
    6:15. In good condition; no symptoms up to this time, excepting a little tendency to droop.
    Following evening. Animal normal.

The venom of the Coral snake (*Elaps fulvius*) is affected to a less degree.

---

[1] Very prolonged boiling, as has been shown by Fayrer and by Ward, lessens greatly, and at last destroys toxicity in cobra venom. The efficient cobra peptone is, as we have seen, converted into a coagulable albuminoid, which is then incapable of destroying life.

Experiment.—Boiled 0.015 gram dried Coral venom dissolved in 1 c. c. distilled water.

    Time of injection?
    5:45. Very weak.
    6:00. Nearly dead.
    6:10. Dead. No local effects. Blood coagulates perfectly.

A smaller amount of venom unboiled kills in from 10–15 minutes with decided local effects.

From the experiments with the venom of the *Crotalus adamanteus* detailed above it appears as though the toxicity of the venom was completely destroyed by boiling, but Weir Mitchell found some years ago that boiling did not destroy the poisonousness of the venom of the *Crotalus durissus*, and further work of our own led us to believe that the want of toxicity of our boiled solutions was only apparent, and that there was accordingly a poisonous principle still present, but not in deadly quantities. We therefore made some further observations, using larger amounts of venom.

Experiment.—Dissolved three drops of fresh *Crotalus adamanteus* venom in 1.5 c. c. distilled water and boiled.

    4:40. Injected into the breast muscles of a pigeon.
    6:10. No positive effects.
    Following morning. Dead, no characteristic local effects.

Experiment.—Dissolved 0.12 gram venom (*Crotalus adamanteus*) in 2 c. c. distilled water and boiled for two or three minutes.

    4:40. Injected the above into two pigeons, giving each half.
    Death within fourteen hours in both pigeons. There was some slight local effect, but nothing comparable to what is observed in the unheated venom. There were no extravasations, and the blood was clotted. The stench from putrefaction at the points of injection was very great, and the muscles around them presented a pale-grayish color as though they had been boiled.

A like result was obtained in the case of another pigeon experimented on in the same way.

From the above series of experiments it is perfectly clear that heating the dissolved venom beyond a definite point, varying no doubt in different venoms, lessens its toxic power. Boiling for some minutes does not destroy the poisonous capacity of the venoms, but simply impairs this quality to a varying degree, depending upon peculiarities in the toxic constituents, as we shall hereafter have reason to observe.

Fayrer and Wall, as already noted, found that *prolonged* boiling of solutions of Cobra venom completely destroyed the poisonous activity of that secretion. We accordingly made some similar experiments with solutions of the venom of the *Crotalus adamanteus* with analogous results.

Experiment.—0.03 gram of the dried venom of the *Crotalus adamanteus* was dissolved in a little distilled water and boiled for ten minutes in a water bath. After being allowed to cool it was injected into the breast of a pigeon.

    1:56. Injection practised.
    1:57. Weak.
    2:00. Convulsions.
    2:37. Since last observation has been lying on its side, very weak.
    2:43. Dead.

EFFECTS OF VARIOUS AGENTS ON VENOM.

In a subsequent experiment the solution of venom was boiled for forty minutes. Three minutes after the injection the pigeon vomited; no other toxic symptoms were observed. In another experiment, in which the venom was boiled for one hour, no symptoms occurred but vomiting. Both of these pigeons were watched for three days, but in neither of them did any poisonous symptoms ensue.

The Effects of Alcohol.—When alcohol is added to fresh venom or to an aqueous solution of venom a copious white precipitate occurs. The following experiments were made to determine if the active principles were entirely precipitated by the alcohol, and if the precipitate was poisonous.

Experiment.—Four drops of the venom of the *Crotalus adamanteus* were placed in 1 c. c. absolute alcohol. The precipitate was filtered and washed with an additional amount of alcohol, the filtrate then being evaporated spontaneously to 1 c. c.

The *precipitate* was placed in 1 c. c. distilled water and injected into the breast muscles of a pigeon at 5:11.

    5:17. Too weak to stand.
    5:21. Dead. There was very little local effect.

The *filtrate* was injected into another pigeon, as above, at 5:22.

    5:26. Vomits; no further effects.

From this experiment it is obvious that the presence of alcohol does not destroy toxicity. Further observations were made to learn the effect of a more prolonged action, and if the precipitate was soluble in water.

Experiment.—0.06 gram of dried *Crotalus adamanteus* was dissolved in 3 minims of distilled water and this was added to 3 c. c. absolute alcohol (Squibb's) causing a dense precipitate. The mixture was allowed to stand for three days. It was then filtered, the precipitate being several times washed with the filtrate and finally with fresh absolute alcohol.

The precipitate was finally washed from the filter by distilled water, allowed to dry, then digested in distilled water for twenty-four hours, and, after being filtered, was washed with distilled water. The filtrate was cloudy, and on being allowed to stand for one and a half hours cleared somewhat, there being an upper layer of clear fluid and some sediment.

One-fourth of the filtrate was now injected into the breast muscles of a pigeon at 4.43.

    4:54. Unable to stand.
    5:10. Dead. There is exceedingly little local effect. The tissues at point of injection are suffused with blood.
    5:45. Blood still fluid.

To one-fourth of the filtrate one minum of acetic acid was added, which caused the mixture to become clear.

    4:41. Injected into a pigeon as above.
    4:52. Rocking.
    4:54. Down.
    5:58. Dead. *There is absolutely no local effect* and there is no *suffusion of blood* in the tissues as in the previous experiment.

To one-fourth of the filtrate a few crystals of sodic chloride were added, which rendered the solution clear.

    4:49. Injected as before.
    4:55. Rocking.
    4:58. Down.
    5:57. Dead. *The local effect is intense;* great blackening and infiltration of fluid blood.

It will have been seen that even after subjection for three days to the action of absolute alcohol the venom has not lost its toxicity. It further appears that the addition of acetic acid or sodic chloride, while rendering the undissolved material soluble delays the time of death, and that the local effects of the poison are destroyed by the acid and intensified by the sodic chloride. The action of the acid is probably due either to a powerful local constricting action on the tissues or else to a modification of the properties of the poison. We can give no reason for the cause in the delay of death after the addition of the sodic chloride. As the animal in this observation lived longer than in the first, the increased local effect may be in this way partially accounted for.

The filtrate becomes very turbid by boiling, and gives a decided precipitate with nitric acid, thus proving that the water has actually dissolved some of the precipitate, and consequently that the toxicity of the filtrate cannot depend merely upon the undissolved particles of precipitate carried through the filter.

It is interesting to learn whether alcohol dissolves any poisonous element of the venom. In one of the above experiments the only effect following the injection of the alcohol filtrate was vomiting, but the objection may be made that the alcohol was in sufficient quantity to act as a physiological antidote to any poisonous element of the venom which it might have contained. We therefore made a further test of this matter by using Cobra venom, which is more powerful than that of the Crotalus, and using it in larger quantities.

Experiment.—Dissolved 0.1 gram Cobra venom in two drops of distilled water, and digested in 2.5 c. c. absolute alcohol for about ten days. The mixture was then filtered, and the filtrate evaporated spontaneously to ⅔ of a c. c. This was injected into a pigeon without any effect.

The following observations with Cobra venom are of great value as throwing light upon the different results obtained by various investigators in studying the action of alcohol on venom. In this series of experiments varying proportions of water were used to dissolve the venom.

Experiment.—Dissolved 0.02 gram dry Cobra venom in *three* drops of distilled water, then added 1 c. c. absolute alcohol and filtered.

    (I.) 4:37. Injected into the breast of a pigeon the above *filtrate*—no symptoms.
    (II.) 4:41. Injected the *precipitate* in a little water.
    4:50. Dead.

Experiment.—Dissolved 0.03 gram Cobra in *ten* drops of distilled water and added 1 c. c. absolute alcohol and filtered.

    (I.) 5:00. Injected the *filtrate* as above.
    5:30. Sick.
    5:53. Unable to stand; extremely feeble.
    5:55. Dead.

(11.) 5:05. Injected the *precipitate* with water.
5:07½. Dead.

In the first series the results are the same as in previous experiments, but in the second series, where a much larger quantity of water was used, the filtrate caused death in fifty-five minutes, thus proving that if sufficient water be present, enough of the poison is carried with the filtrate to cause death, notwithstanding the larger amount of alcohol present and its attributed antidotal action.

The Action of Absolute Alcohol upon the Dried Venom.—If dried venom be placed in *absolute* alcohol and the mixture allowed to stand for some time, even for months, it will be found that the venom undergoes no change in its poisonous activity, nor does it appear that the alcohol dissolves out any of the poisonous principles, since it is found to be innocuous after injection, and does not give any reaction for proteids.

The Effect of the Caustic Alkalies on the Toxicity of Venoms. Caustic Potash.—When caustic potash is added to a solution of venom the latter becomes perfectly clear. If the quantity of salt added to the solution is below a definite limit no decided alteration in the capacity to kill is noticed, but as this quantity increases obvious results are observed, first a diminution in the activity of the poison, and at last a complete loss of toxicity.

Experiment.—Dissolved 0.03 gram dried *Crotalus adamanteus* venom in 1 c. c. of distilled water in which was previously dissolved 0.0037 gram potassic hydrate.

3:45. Injected into the breast of a pigeon.
4:48. Weak.
5:00. Unable to walk; 6:00 ditto.
6:30. Dead. Heart arrested in systole; no ecchymoses; well-marked local effect; blood fluid at the end of sixteen hours.

Experiment.—The same as above, using 0.0075 gram potassic hydrate.

3:43. Injected.
5:00. Weak.
6:00. Weaker; slight local effect.
Following morning. Animal living, but weak; the local effect is well marked.

Experiment.—The same, using 0.015 gram potassic hydrate.

3:47. Injected.
7:00. No symptoms up to this time.
7:30. Sickish.
Following morning. Sickish; some slight local effect at point of injection.

Experiment.—The same, using 0.03 gram potassic hydrate.

3:50. Injected.
7:00. No symptoms up to this time.
Following morning. No symptoms; no local effects.

This experiment was repeated in two other pigeons with a like result.

The last series of experiments prove clearly that the addition of potassic hydrate to a solution of venom, if in sufficient quantity, produces a decided effect on the activity of venom, and that if added to the venom of the *Crotalus adamanteus* in a quantity equal to the weight of the dried poison the lethal action is entirely destroyed. In one experiment made with the venom of the *Crotalus horridus* the same holds good, but our experiments with Cobra venom show that a larger proportional quantity is needed to destroy its power.

Experiment.—Dissolved 0.015 gram Cobra venom in 0.5 c. c. distilled water, then added an equal amount of potassic hydrate.

    4:05. Injected into the breast muscles of a pigeon.
    4:22. Unable to stand.
    4:32. Convulsions.
    4:37. Dead.

This experiment was repeated with a similar result.

A larger proportion of the potassic hydrate was used in the following observations:—

Experiment.—The same as above, using 0.03 gram (double the amount) of potassic hydrate.

    4:51. Injected into the breast of a pigeon.
    No effects.

Repeated this experiment with a similar result.

In one instance, however, we found that 0.06 gram potassic hydrate did not effectually counteract the poisonous activity of 0.015 gram dried Cobra.

It has been suggested that the non-poisonous action of venom treated with potassic hydrate and injected hypodermically, as in the above experiments, depends upon an effect of the potassic salt on the tissues, causing a considerable delay in the absorption of the poison, and this suggestion seems strengthened by the result in a rabbit of an *intravenous* injection of 0.015 gram Crotalus venom with 0.06 gram potassic hydrate in 1 c. c. distilled water. The animal became very sick soon after the injection, which was given in the evening, and remained in this condition at the end of an hour, when the observation ceased. The following morning it was found dead, with post-mortem appearances of the effects of venom. Also, in another animal, which was given intravenously 0.015 gram of venom with a similar amount of potassic hydrate, death occurred as promptly as with pure venom; in fact rather earlier.

In another set of experiments on pigeons, we carefully neutralized the potassic hydrate before injecting. We used in all of this series sulphuric acid as the neutralizing principle, so that a harmless potassic sulphate was formed. The results of this group of experiments also go to show that the potassic hydrate prevents the absorption of the venom.

Experiment.—Dissolved 0.015 gram dried venom of the *Crotalus adamanteus* in 1 c. c. distilled water and added 0.015 gram potassic hydrate, then carefully neutralized with acetic acid.

This was injected into the breast of a pigeon, causing death in sixteen minutes.

Experiment.—Dissolved 0.03 gram dried *Crotalus adamanteus* venom in 1 c. c. distilled water and added 0.015 gram potassic hydrate, and then neutralized as above.

    4:19. Injected as above.
    4:55. Weak; breathing rapid.
    6:20. Much weaker.
    Following morning. Dead. Decided local effect; blood fluid and dark.

Experiment.—The same, only using 0.0075 potassic hydrate.
 4:53. Injected as before.
 5:00. Weak; breathing deep.
 5:10. Dying.
 5:19. Dead. Slight local effect; blood fluid and dark.

The records of the above experiments, which are in accord with Wall's, show that the results after the addition of the potassic hydrate are not the same as in the series where the alkali was not neutralized, thus proving that the effect of the action of the added alkali does not remain after the latter is neutralized.

In previous observations we found that solutions of venom were more or less impaired by boiling, and that this was particularly marked with the venom of the *Crotalus adamanteus*, 0.015 gram being rendered completely innocuous to pigeons. It was afterwards found that no coagula were formed by heating solutions of venom to which had been added some potassic hydrate, as in the above experiments. This led us to study the results of heating solutions of venom to which the potassic hydrate was added to learn if heat was capable of destroying or impairing toxicity without the occurrence of coagulation as a necessary event.

Experiment.—Dissolved 0.015 gram of the venom of the *Crotalus adamanteus* in 1 c. c. distilled water and added 0.015 gram potassic hydrate, and subjected the solution, as in previous experiments, to a gradually increasing temperature up to 74° C. It was then injected into a pigeon. At the end of twenty-four hours there was no effect.

In this experiment the temperature to which the solution of venom was submitted was below the point at which serious impairment of the poisonous power of the venom occurs, yet the amount of potassic hydrate was sufficient to destroy its action. Other experiments were made in which the quantity of potassic hydrate was not sufficient to effect this end. We found in previous experiments that 0.0037 gram potassic hydrate was not sufficient to destroy the toxicity of 0.03 gram of *Crotalus adamanteus* venom, although the time of the occurrence of death was considerably delayed.

We used similar amounts of venom and alkali in the three following experiments, using 0.5 c. c. distilled water for the solutions.

Experiment.—Dissolved 0.09 gram of *Crotalus adamanteus* venom in 1.5 c. c. distilled water and added 0.011 gram of potassic hydrate. This solution was divided into three parts. One of which was heated to 76.5° C., one to 79.5° C., and the other to 83.5° C. Each of which was injected into the breast of a pigeon and without any evil consequence following within twelve hours.

These results indicate that heat impairs the poisonous activity of venom under the above conditions, even though coagulation does not occur. In previous experiments recorded it was found that at a temperature of 79.5° 0.03 gram of *Crotalus adamanteus* venom was rendered non-toxic. The explanation of the further impairment of the action of the poison by heating its solutions having potassic hydrate dissolved in them lies probably in the fact that the potassic hydrate is placed by heat under condition of greater activity. The non-coagulability of solutions of venom to which potassic hydrate was added is no doubt due to the alteration of the coagulable proteids into alkali-albumins, and as a moderate degree of heat increases the rapidity of this change, it is possible that the smaller amount of alkali is as effective under these conditions as the larger amounts under ordinary conditions. It is not at all improbable that the prolonged action of potassium hydrate on solutions of venom may convert all of the globulins into alkali-albumins and thus destroy their poisonous activity.

Sodic Hydrate.—The effect of sodic hydrate on solutions of the venoms of the *Crotalus adamanteus* and *horridus* appears to be the same as that of the potassic salt. In one experiment with the *Crotalus adamanteus*, using equal quantities (0.03 gram) of the dried venom and alkali, no poisonous effects followed its injection; and in another experiment in which 0.015 gram of venom and 0.007 gram sodic hydrate were used the animal was rendered somewhat sick, but fully recovered.

In one experiment with the venom of the *Crotalus horridus*, using equal quantities (0.015 gram) of the venom and sodic hydrate, no poisonous symptoms followed.

The effect on solutions of dry Cobra venom, as in the case of the potassic salt, is not so marked.

Experiment.—Dissolved 0.015 gram dry Cobra venom in 0.5 c. c. distilled water and added 0.015 gram sodic hydrate.

4:08. Injected into the breast of a pigeon.
4:15. Unable to stand.
4:27. Dead.

In two other experiments, using double the quantity of sodic hydrate, one animal died in one hour, and the other in a little less than three hours. Double amounts therefore decidedly impair toxicity. In another experiment, in which four times the quantity of sodic hydrate was used (0.015 gram dried venom + 0.06 gram NaHO), no poisonous symptoms followed.[1]

The Effects of Ammonia.—The dry venom of the *Crotalus adamanteus*, which was the only one used, forms with aqua ammonia a turbid solution, such as is formed with water. The effect on the toxicity of the venom exerted by the ammonia is not so marked as with the potassic or sodic hydrates.

Experiment.—Dissolved 0.03 gram dried venom in *two minims* aqua ammonia (20°) with 1 c. c. distilled water.

5:29. Injected into the breast muscles of a pigeon.
5:37. Unable to walk.
5:46. Convulsions; death. The local lesions are decidedly lessened by the alkali.

Experiment.—The same, using *six minims* aqua ammonia.

4:14. Injected as above.
6:00. No marked symptoms up to this time, excepting droopiness. The local effect is slightly more marked than in No. 1.
Following morning the animal was dead.

In three other experiments in which eight minims of aqua ammonia were used two of the animals were found dead the following morning and one recovered. In

---

[1] See Shortt, Wall. op. cit., p. 133. On the effects of alkalies and of permanganates, see Vincent Richards, F.R.C.S. Ed., etc., op. cit.

another experiment in which the alkali was neutralized by sulphuric acid, death did not occur for *four* hours.

In the first experiment, in which a very small amount of ammonia was used, death occurred in less than twenty minutes; in the next, in which three times the quantity of ammonia was used, death did not ensue for some hours, while in the next three a more positive effect was no doubt apparent in the fact that one of the pigeons recovered. In the last experiment death did not occur for over four hours, even after neutralization of the alkali, indicating, as in the case of the potassic hydrate, that some permanent effect had been exerted on the venom by the ammonia.

Potassium Carbonate.—Two experiments made with the venom of the *Crotalus adamanteus* render it probable that the potassic carbonate does not exert any decided effect.

Experiment.—Dissolved 0.015 gram venom in 1 c. c. distilled water and added 0.015 gram potassic carbonate.

  4:16. Injected into the breast of a pigeon.
  4:22. Down.
  4:25. Dead. No appreciable local effect.

Experiment.—Dissolved 0.03 gram venom in 1 c. c. distilled water and added 0.12 gram potassic carbonate.

  5:25. Injected into the breast muscles of a pigeon.
  5:45. Down; observation now ceased.
  Following morning found dead; slight local effect.

Nitric Acid.—The powerful destructive action exerted by this acid on albuminoids suggests at once that it would in all likelihood completely destroy the poisonous properties of venom, yet it has been asserted that such is not the case. In the latter instance the result was no doubt due to the insufficiency of acid used, as we have clearly determined in our experiments.

Experiment.—Dissolved 0.03 gram *Crotalus adamanteus* venom in 0.5 c. c. distilled water and added 2½ minims C. P. nitric acid, which caused a considerable precipitate.

  3:32. Injected the above into the breast of a pigeon.
  3:33. Convulsions, followed by death.

From this result it seemed probable that not enough acid had been added to throw down all of the precipitable proteids. In another experiment the acid was added to a solution of venom and the mixture filtered. The filtrate was now tested with nitric acid and a further precipitate occurred. This process was repeated until no further precipitate followed. The filtrate was set aside, and the precipitate on the filter washed with dilute nitric acid and then with water.

Experiment.—5:05 injected into the breast of a pigeon the above *filtrate*, which measured 3 c. c. and contained 1 c. c. nitric acid.

  6:05. No symptoms except slight droopiness.
  Following morning no effects from venom.
  5 April, 1886.

Experiment.—5:55 injected the *precipitate* in 1 c. c. dilute nitric acid with which it had been in contact for two hours.

 6:05. No symptoms.
 Following morning animal in good condition.

It will thus be observed that the acid has completely destroyed the toxicity of venom. We made still another experiment in which the venom was rubbed up in a mortar and the acid added to it, and then diluted with water.

Experiment.—0.03 gram dried Crotalus venom was rubbed in a mortar until powdered, and 4 gtt. C. P. nitric acid added. This formed a pasty mass of an orange-yellow color. With 1 c. c. distilled water it formed a cloudy, orange-yellow solution.

The above was injected into the flank of a half-grown rabbit, without any symptoms of venom poisoning following within twelve hours.

A similar experiment was made with a pigeon with a like result. The acid, however, having been neutralized with sodic carbonate before injection.

Muriatic Acid.—This acid does not seem to exert so strong an effect. Only one experiment was made.

Experiment.—0.015 gram dried Crotalus venom was rubbed in a mortar, and to it was added 4 gtt. C. P. muriatic acid forming a clear solution. With 1 c. c. distilled water it made a turbid solution.

 3:44. Injected the above into the breast muscles of a pigeon.
 5:00. Very sick.
 5:50. Nearly dead.
 Following morning dead; no local lesions from venom.

Here the amount of venom used was only one-half of that employed in the nitric acid experiment. The quantity of acid was the same, but in this experiment a pigeon was used.

As in the series with nitric acid, an experiment was also made in which the dried venom was powdered in a mortar and a few drops of the pure acid used. About 1 c. c. of distilled water was added, and the mixture neutralized with sodic carbonate. It was then injected into the breast of a pigeon with the result of death in twenty-six minutes.

Sulphuric Acid.—Repeated the above, using instead of the muriatic acid 5 gtt. sulphuric acid. The venom and acid formed a clear syrupy solution which became milky by the addition of the water.

 3:53. Injected as above.
 5:50. Sickish.
 Following morning dead; no local symptoms of venom poisoning.

Dr. Mitchell had observed that if the acid was afterwards neutralized the action of the venom was not affected. The delay of death in this experiment seems to be due to the action of the non-neutralized acid. We, however, made an experiment by powdering the dried venom (0.015 gram) in a mortar, adding a few drops of the pure acid, diluting then with about 1 c. c. distilled water, and neutralizing with sodic carbonate. This was injected into the breast of a pigeon.

For some time after the injection the bird was weak, and continued in a feeble condition until eighteen hours after the injection, when death ensued.

It seems quite remarkable that such a powerful acid as sulphuric does not completely destroy the poisonous properties of the venom, and it is even more curious that pure muriatic acid seems to be without effect.

Acetic Acid. Experiment.—Dissolved 0.02 dried venom (*Crotalus adamanteus*) in 0.1 c. c. distilled water and added 3 minims of glacial acetic acid.

    4:30. Injected into the breast of a pigeon.
    4:37. Incoördination.
    4:41. Dead.

Death occurred in this experiment in such a short time that it was thought that the acid itself might have contributed to this end. We therefore made another experiment in which the acid was neutralized.

Experiment.—Prepared the venom as before, only neutralizing the solution with sodic carbonate.

    4:35. Injected into the breast of a pigeon.
    5:10. Pigeon unable to stand.
    5:15. Dead.

The result of this experiment indicates that the presence of the free acid aids the toxic action of venom.

Hydrobromic Acid. Experiment.—Powdered 0.015 gram dried *Crotalus adamanteus* venom in a mortar and added 5 gtt. hydrobromic acid (sp. gr. 1.274), after five minutes added 0.5 c. c. distilled water. The venom and acid formed a slightly reddish-colored solution, which became milky when diluted with water.

    4:25. Injected into the breast muscles of a pigeon.
    4:45. Sickish.
    4:55. Unable to stand. (Final result not noted, but death most certainly followed.)

We repeated the above experiment, using 10 gtt. of acid mixed with an equal part of water, before dissolving the venom in it.

    2:49. Injected as above.
    3:07. Rocking.
    3:30. Dead; local effects of the venom apparent.

Notwithstanding we used double the amount of acid in this experiment, it does not appear as though the activity of the venom was made to differ much from that noted in the previous experiment. Since the previous dilution of the acid before mixing with the venom might have affected its action a third experiment was made in which the same quantity of acid was added, without dilution, to the powdered venom.

Experiment.—Powdered 0.015 gram dried venom and added 10 gtt. hydrobromic acid, then 1 c. c. distilled water.

    4:20. Injected the above into the breast muscles of a pigeon.
    5:00. No apparent effect.
    5:10. Sickish.
    6:00. Sickish.
    Following evening. Well.

This last experiment was repeated with the modification of leaving the acid in contact with the venom for one-half hour before the addition of the water. It was then injected as above without any obvious effects following.

The destructive action of the acid on the venom of the *Crotalus horridus* seems to be the same if we can judge from the single experiment which follows.

Experiment.—Powdered 0.015 gram dried venom and added 10 gtt. hydrobromic acid, which formed a muddy solution with a reddish color.

    5:18. Injected into the breast of a pigeon without any obvious effects within twenty-four hours.

The effect on the activity of Cobra venom, using similar quantities of venom and acid is very different.

Experiment.—Repeated the above, only substituting Cobra venom.

    4:48. Injected into the breast muscles of a pigeon.
    5:18. Sick; breathing difficult.
    5:30. Breathing more difficult; convulsive movements; incoördination.
    5:35. Dead.

Tannic Acid.—The action of tannic acid upon albuminoids is so decided that we might confidently expect, since we find the poisonous elements in venoms to be proteids, that the activity of venom would be greatly diminished or entirely destroyed by it. In one experiment made with the venom of the *Crotalus adamanteus* we found comparatively little effect.

Experiment.—Dissolved 0.03 gram dried venom in a little distilled water and added 1.5 c. c. saturated solution of tannic acid.

    3:35. Injected into the breast muscles of a pigeon.
    4:00. Droopy.
    4:45. The same. Following morning dead.

It will be observed that there is a great delay in the action of the venom, possibly due to the powerful local constrictive action of the tannic acid on the tissues, and possibly, also, to a direct action of the acid on the venom itself. As death may have resulted from the tannic acid we made a control experiment in which 1.5 c. c. saturated solution was injected into the breast of a pigeon. The animal did not exhibit any signs of active poisoning, but it died at the end of the fourth day.

Alum.—We made but two experiments with alum, one with the venom of the *Crotalus horridus* and one with *Cobra.*

Experiment.—Dissolved 0.015 gram dried venom in 0.5 c. c. distilled water and added 3 gtt. saturated solution of alum (18° C.), but no precipitate occurred; we then gradually added powdered alum nearly to saturation, which caused precipitation. The precipitate was filtered off, and the clear filtrate tested by the further addition of alum to see if any more precipitation would occur, with a negative result. The precipitate and filtrate were now mixed together and injected into the breast of a pigeon without any poisonous result occurring within forty-eight hours.

In another experiment in which 0.06 gram of dried Crotalus venom was used, the animal died in forty-five minutes.

EFFECTS OF VARIOUS AGENTS ON VENOM.

Alum added to saturation does not precipitate the peptone, although it precipitates all of the coagulable proteids.

The following is the experiment with Cobra venom:—

Experiment.—Dissolved 0.015 gram dried venom in 0.5 c. c. distilled water and added alum to saturation (16° C.).

- 4:32. Injected into the breast muscles of a pigeon.
- 4:50. Down.
- 4:52. Dead.

This last experiment is of interest in proving that even so powerful an astringent as alum is not sufficiently strong to prevent the prompt absorption of the poison. Death followed in twenty minutes.

Chlorine Water.—This reagent does not seem to exert any influence.

Experiment.—Dissolved 0.015 gram *Crotalus adamanteus* venom in 0.5 c. c. distilled water and added 0.5 c. c. fresh chlorine water.

- 4:28. Injected into the breast muscles of a pigeon.
- 4:52. Down.
- 5:10. Dead.

Bromine.—The action of bromine in bromohydric acid solution is very marked.

Experiment.—Powdered 0.015 gram dried *Crotalus adamanteus* venom in a mortar and added 2 gtt. of bromine in 4 or 5 gtt. bromohydric acid, then added 0.5 c. c. alcohol.

- 5:05. Injected into the breast of a pigeon.
- 5:30. No effect.
- Twenty-four hours. No effect.

This experiment was repeated once with Crotalus venom and once with Cobra, using water as the diluent instead of alcohol. In both experiments we found a similar result, thus proving that the activity of the venom is completely destroyed by this reagent.

Iodine. Experiment.—Dissolved 0.015 gram dried venom of *Crotalus adamanteus* in 0.33 c. c. distilled water, then added 0.5 c. c. tr. iodine which formed a dense brown precipitate.

- 5:07. Injected into the breast of a pigeon.
- No poisonous effects within twenty-four hours.

If, however, the amount of iodine be much smaller the venom is still potent, as is shown by the following experiment.

Experiment.—Dissolved the venom as above, then added 1 drop tr. iodine and afterwards 1 c. c. distilled water.

- 4:56. Injected into the breast of a pigeon.
- 5:05. Weak.
- 5:15. Dying.

Iodine + Potassic Iodide. Experiment.—Dissolved 0.015 gram dried venom of *Crotalus adamanteus* in 0.5 c. c. distilled water, then added a saturated solution of equal parts of tr. iodine and potassic iodide.

- 4:41. Injected the above into the flank of a small rabbit (half grown).
- The animal died in about eighteen hours.

The delay in the occurrence of death in this experiment was considerable, and that this was due to the action of the iodine on the venom is rendered probable by the results of the preceding experiments with iodine and by the following experiment with the potassic iodide.

Potassic Iodide.—This salt does not seem to exert any influence upon the activity of venom.

Experiment.—Dissolved 0.015 gram dried venom of the *Crotalus adamanteus* in 1 c. c. saturated solution of potassic iodine.

 4:31. Injected into the breast of a pigeon.
 4:40. Down.
 4:45. Dead.

Potassic Bichromate. Experiment.—Dissolved 0.03 gram dried *Crotalus adamanteus* venom in 1 c. c. distilled water and added 0.01 gram potassic bichromate.

 4:14. Injected into the breast of a pigeon.
 4:20. Down.
 4:25. Convulsions followed by death.

Experiment.—Dissolved 0.004 gram dried venom in 0.5 c. c. distilled water and added 0.03 gram potassic bichromate dissolved in 0.33 distilled water, which produced a dense coagulum.

 3:38. Injected into the breast muscles of a pigeon.
 4:05. Dead.

Potassic Permanganate. Experiment.—Dissolved 0.03 gram dried *Crotalus adamanteus* venom in 0.5 c. c. distilled water and added 0.06 gram permanganate in 0.5 c. c. distilled water. This formed a very cloudy solution.

 5:27. Injected into the breast of a pigeon. Death occurred within forty-eight hours.

Experiment.—The same, using 0.015 gram of the permanganate. At the end of the second day no poisonous effects from the venom. The parts where the injection was made look as though they would slough.

Experiment.—The same, using 0.005 gram of the permanganate. The solution formed is a dark wine color.

 4:37. Injected into the breast of a pigeon without effect.

Experiment.—The same, using 0.0038 gram of permanganate.

 4:26. Injected as above.
 4:42. Down.
 4:45. Dead.

Experiment.—The same, using 0.0025 gram of permanganate.

 3:57. Injected as above.
 4:06. Down.
 4:10. Dead.

In another experiment, the mixture was injected into the femoral vein of a rabbit, using 0.005 gram permanganate. The animal lived, and at the end of the second day was apparently unaffected.

In one observation made with the venom of the *Crotalus horridus* 0.015 gram of venom was dissolved in 0.5 c. c. distilled water, to which was afterwards added 0.008 gram of the permanganate. After standing for twenty-four hours the mixture was very thick and tarry, and would not flow from the inverted test-tube. It seems from this that the full extent of the action of the permanganate on the venom is not exerted for some hours.

The permanganate is efficient in destroying the activity of Cobra venom.

Experiment.—Dissolved 0.015 gram dried Cobra venom in 0.5 c. c. distilled water and added 0.015 gram permanganate.

 4:35. Injected into the breast muscles of a pigeon. No symptoms of venom poisoning within twenty-four hours.

Peroxide of Hydrogen.—Notwithstanding the powerful nature of the peroxide of hydrogen as an oxidizer, it does not seem to affect to any great extent the poisonous activity of venom. Only one experiment was made.

Experiment.—Added 3 drops of fresh venom of the *Crotalus adamanteus* to 3 c. c. fresh solution of peroxide of hydrogen, specially prepared by Prof. Leeds, of Hoboken.

 5:05. Injected into the breast muscles of a pigeon.
 5:15. Unable to stand; decided local effects appearing.
 6:05. Dead, with all the usual phenomena of venom poisoning.

The quantity of peroxide of hydrogen used in this experiment was so large that the test was a satisfactory one.

Argentic Nitrate.—Notwithstanding the powerful action of nitrate of silver on albuminoids it does not seem to possess great power to disturb the toxicity of venom.

Experiment.—Dissolved 0.015 gram of dried venom of *Crotalus adamanteus* in 3 c. c. distilled water, to which was afterwards added 0.015 gram nitrate of silver, forming a decidedly milky solution.

 4:40. Injected into the breast of a pigeon.
 4:50. Down; deep breathing, gasping.
 4:53. Dead.

As there was a possibility of the quantity of salt being insufficient for the amount of venom, another experiment was made in which double the weight of nitrate was used. The mixture was injected into the breast of a pigeon. At the end of three days no symptoms of venom poisoning had occurred.

Mercuric Chloride.—When mercuric chloride is added to a solution of Crotalus or Moccasin venom a dense precipitate occurs, consisting of all the proteids in solution. In order to learn if the precipitated proteids still retained any toxic power we dissolved 0.03 gram of dried venom of the *Crotalus adamanteus* in 1 c. c. distilled water and then added 0.03 gram mercuric chloride. The precipitate was collected on a filter and repeatedly washed with distilled water. During this washing the precipitate seemed to diminish a little in quantity, and was no doubt partially dissolved.

 3:30. The precipitate in 1 c. c. distilled water was injected into the breast of a pigeon.
 6:00. No symptoms up to this time.
 Twenty-four hours—the animal showed no signs of venom poisoning.

Ferrous Sulphate.—Three experiments were made with the sulphate of iron with results materially different; the difference no doubt depending upon the mode of administration. In all the same quantities of venom and salt were used, but in one the solution was injected simply beneath the skin and in the others directly into the muscles of the breast. In the former the animal did not die until after the lapse of nearly thirty-six hours, while one of the others died remarkably soon —within four minutes after the injection, and the third in twenty-eight minutes.

Experiment.—Dissolved 0.03 gram dried venom of the *Crotalus adamanteus* in 1 c. c. distilled water and then added 0.03 gram ferrous sulphate. The addition of the iron salt renders the solution clear.

    3:40. Injected *beneath the skin* of the thigh of a pigeon.
    6:00. No apparent effects.
    Twenty-four hours—no effects. Thirty-six hours—dead. Slight local effects of venom, but the destructive action of the iron salt on the tissues is much more prominent.

Experiment.—The same as above.

    3:32. Injected *into the breast muscles* of a pigeon.
    3:36. Convulsions; death. No local lesions.

In another experiment the bird died in twenty-eight minutes after injection.

It must be concluded from this that the ferrous sulphate does not destroy the activity of the venom.

Dialyzed Iron.—When dialyzed iron is added to a solution of venom all of the proteid matter is precipitated, and the filtrate is found to give no reaction for proteids with the xanthoproteic or picric-acid tests. The precipitate is brown, and so gelatinous that if the solutions are somewhat concentrated it does not flow. The precipitate does not dissolve in distilled water, yet it must be very soluble in the tissues since the toxic effects of the venom rapidly appear after its injection. We made two experiments, both with Moccasin venom, one with the dried and the other with fresh venom.

Experiment.—Dissolved 0.015 gram dried Moccasin venom in 0.5 c. c. distilled water and added 3 gtt. dialyzed iron. This caused a considerable amount of brownish gelatinous precipitate which thickened the mixture appreciably. Now added 1 c. c. distilled water.

    3:20. Injected into the breast muscles of a pigeon.
    3:25. Down.
    3:45. Dead.

Experiment.—Took two drops of fresh Moccasin venom and added first 5 gtt. dialyzed iron, and then 1 c. c. distilled water. The iron and venom made a very thick brownish mixture.

    5:18. Injected into the breast muscles of a pigeon.
    5:30. Dead.

One experiment was made in this connection to see if dialyzed iron exerted any poisonous effect, we injected thirty drops into the breast muscles of a pigeon, without toxic result.

Ferric Chloride.—We have used the chloride of iron in two forms; the officinal *tincture*, U. S. P., and the officinal *liquor*. Both these solutions greatly affect the poisonous activity of venom, the latter, indeed, if used in sufficient quantity, wholly prevents the occurrence of any of the symptoms of venom poisoning. The tincture does not appear to be nearly as efficient.

Experiment.—Dissolved 0.015 gram dried venom of the *Crotalus adamanteus* in 0.5 c. c. distilled water and added 10 gtt. *tr.* chloride of iron. As the iron was added the solution cleared, but in a few moments became milky, and finally thick with whitish precipitate.

 4:22. Injected into the breast of a pigeon.
 5:00. No symptoms.
 5:45. No symptoms.
 Following morning—dead. No local effect.

In two similar experiments, in which double the quantity of the tincture of iron was used, the result was much the same, the time of death being notably delayed. The following experiments were made with the *liquor*:—

Experiment.—Dissolved 0.015 gram dried venom of *Crotalus adamanteus* in 0.5 c. c. distilled water and added 4 gtt. *liquor ferri chloridi*. A heavy precipitate fell.

 4:45. Injected into the breast of a pigeon.
 5:00. Very quiet.
 6:00. No symptoms, and none of venom poisoning within two days.

A similar experiment was made with identical results.

In one experiment with the venom of the *Crotalus horridus*, in which only two drops of the liquor were used, the animal showed no evidences of poisoning; and in four experiments made with the dried venom of the *Moccasin*, in which 0.015 gram of dried venom was used and eight, four, two, and one drop of the liquor were used, three animals gave no symptoms of venom poisoning, and one died on the third day—the animal receiving the injection containing but one drop of the iron. This was the only pigeon of the four which gave any signs of poisoning. In three-fourths of an hour the bird was shaky, and at the end of three hours decidedly feeble, remaining pretty much in this condition until death.

About the point of injection the iron produced considerable hardening of the tissues.

The effect on Cobra venom is not marked, although in one experiment there was an appreciable delay in the occurrence of death; but in the other, in which the quantity of iron was larger, death occurred with remarkable rapidity.

Experiment.—Dissolved 0.015 gram dried Cobra venom in 0.5 c. c. distilled water and added 2 drops *liquor ferri chlor.* A slight precipitate occurred in the solution after a few moments.

 3:47. Injected into the breast muscles of a pigeon.
 4:15. Convulsions.
 4:27. Dead.

Experiment.—Dissolved 0.015 gram dried Cobra venom in 1 c. c. distilled water and added 5 gtt. sol. perchloride of iron.

This was injected into the breast of a pigeon, with the result of death in twenty seconds.

The reason why ferric chloride is inefficient in destroying the toxicity of Cobra venom no doubt lies in the fact of its main poisonous substance being a peptone, and, like that element in all the venoms, unaffected by the iron, while the principal toxic effects of the Crotalus and Ancistrodon venoms is due to the globulins which are precipitated and chemically altered by the iron salt.

Filtration through Various Substances.—Filtration through *alumina* or *wood charcoal* does not affect the poisonous activity of the venom, but by filtration through *animal charcoal* all of the poisonous material in venom is left behind and the filtrate is accordingly innocuous.

Experiment.—Dissolved 0.03 gram dried Moccasin venom in 2 c. c. distilled water and filtered four times through animal charcoal. The filtrate gives no proteid reaction.

> 4:20. Injected 1.5 c. c. of the filtrate into the breast of a pigeon. At the end of twenty-four hours no symptoms of venom poisoning had occurred, but there was some œdema at the point of injection.

Repeated the above experiment, using 0.045 gram of Moccasin venom, and with similar results.

Snake Bile.—Among the curious substances which have been extolled as antidotes for venom poisoning is snake bile. We made but one experiment, which speaks volumes.

Experiment.—Mixed 1½ minims of fresh venom from a dead *Crotalus adamanteus* with 1 c. c. of bile from the same animal.

> 4:47. Injected into the breast of a pigeon.
> 4:47½. Incoördination.
> 4:50. Gasping respiration.
> 4:55. Convulsive movements.
> 4:56. Dead.

Digestion.—By digestion in strong artificial *gastric* juice made from the pig's stomach the toxic power of venom (Crotalus) is completely destroyed.

Experiment.—Three drops of the glycerine solution of venom (*Crotalus horridus*) (1862) were digested for sixteen hours in about 1 c. c. fresh artificial gastric juice from the pig's stomach.

> 8:30 A. M. Injected into the breast muscles of a pigeon. Up to the end of forty-eight hours no poisonous symptoms ensued.

This experiment was repeated with an identical result. We also made two experiments in which the digestive process was not carried on for such a length of time, in both only four hours, and with similar results. In one we used six drops of the glycerine solution of venom as above—just double the dose—and in the other 0.015 gram of the dried *Crotalus adamanteus* venom.

The results of digestion in artificial pancreatic juice are similar. We made but one experiment, and that with the venom of the *Crotalus adamanteus*.

Experiment.—Digested 0.03 gram dried venom in 1 c. c. freshly prepared pancreatic juice from the pig for twenty-four hours.

> 5:44. Injected into the breast muscles of a pigeon.
> 5:45. Slightly droopy.
> Following morning, no effects apparent.

Résumé.—The above experiments, with others too numerous for detail, have enabled us to confirm Lacerda's and Vincent Richard's views as to the power of permanganate of potassium to destroy venoms. As a local antidote it is for all snake poisons the best.

It is also clear from what we have seen that ferric chloride is a very efficient local destroyer of the venom of our own snakes, which owe their vigor to venom-globulin, but has little value as a local antidote to the peptone which gives power to the poison of the Cobra. The chloride needs to be locally used in full doses, whence it is that the strong liquor ferri chloridi (U. S. P.) is more efficient than the tincture.

That bromine may prove valuable as a local means of relief seems to be plain from our experiments, and is in fact one of their most interesting results. It was used, as we have seen, either as hydrobromic or bromohydric acid. Probably any solution of bromine would answer, and—as was shown by its free local use to control gangrene during our civil war—there need be no fear in using it with freedom.

It has long been known in India that the strong alkalies destroy venom, and this we are able to confirm. Brainerd long ago taught that iodine has destructive value as regards Crotalus venom, and this also seems to us to be true.

In fact many agents more or less alter venoms if allowed to remain long in contact with them, and usually act with increased vigor as the temperature is raised above that of the air; but it is chemically singular that brief exposure of venoms to strong acids should so little affect the toxicity of the poisons in question. Except where otherwise distinctly stated, the chemicals used by us have been added to the poison immediately before injecting it. Enough has been here proved to make it now worth while to study still more carefully the value of bromine and ferric chloride as local poison destroyers. One agent may be at hand or available when others are not, and the more numerous are the means we possess as local antidotes the better is the chance of escape or relief for persons bitten.

CHAPTER IV.

THE EFFECTS OF VENOM WHEN APPLIED TO MUCOUS OR SEROUS SURFACES.

The Effects of Venom when applied to Mucous Surfaces.—The question of the absorption of venom by mucous surfaces is one of great interest, and the verdict of all observers in connection with the venom of the Crotalidæ is that uninjured mucous surfaces, except in the lungs, cannot absorb venom, at least in sufficient quantities to produce death. In experiments with the venom of the Cobra other investigators have gotten results which are directly contrary, but in our own researches a large proportion of the animals survived.

In seven experiments made on pigeons, in which a solution of Cobra venom was placed in the craw by means of an œsophageal tube, six showed no evidences of poisoning and one died. In these experiments the œsophageal tube, which was a small catheter, was oiled and passed with great care into the crop, the solution of venom was then poured through the tube by means of a funnel, and afterwards washed down with a little water.

Experiment.—Dissolved 0.025 gram dried Cobra (*Naja trip.*) in about 1 c. c. distilled water, and placed it in the crop of a pigeon by means of an œsophageal tube. Up to four days the animal showed no signs of poisoning.

Five other experiments, like the above, gave identical results. In one experiment the animal died within twelve hours.

Experiment.—Dissolved 0.013 gram dried Cobra (*Naja trip.*) in about 1 c. c. distilled water and gave to a pigeon, as above, at 4:00 P. M. A little while after the dose the pigeon appeared sickish and remained in much the same condition for about two hours, when observation temporarily ceased. At 8:30 the following morning the bird was dead. The heart was found in systole and contained dark clots. The blood was everywhere coagulated. No apparent lesions were present in any part of the body, and a most careful examination of the mouth, gullet, and crop revealed no abrasions or other raw surface. The crop contained a little cracked corn and a small amount of yellowish fluid.

In another pigeon, etherized, an opening was made through the skin into the craw, and its contents washed out. The pigeon was kept on its back and the edges of the wound were held up by retractors. A solution of venom was placed in the *cul-de-sac* on the left side and the animal watched. In a half hour the bird had convulsive seizures, and at the end of forty minutes was dead. At that time there seemed to be about the same quantity of venom solution in the crop as before. It was, however, somewhat glutinous and darker in color. The mucous membrane preserved its natural tint. On the side in which was the poison the mucous membrane was wrinkled and raised in points like the surface of a mulberry. By stretching the mucous membrane this roughness disappeared. After death it increased somewhat. There was no œdema.

Experiment.—0.01 gram of dried Cobra dissolved in a little water was given to a frog by an œsophageal tube as in the case of the pigeons. The frog presented no toxic symptoms for two hours. After twelve hours it was dead.

Autopsy.—All the tissues had a cyanotic appearance and the animal was perfectly flaccid. The heart was still irritable as well as the intestines. The stomach contained a viscid mass of mucus, which was not bloody, and which was expelled from the stomach by the normal contractility when the organ was cut. A most careful examination of the mouth, gullet, and mucous membrane of the stomach did not reveal any abrasions or other raw surface. The liver seemed pale and decidedly friable.

In Dr. Mitchell's former experiments made in the opened crop, fatal results did not occur with use of fresh or dry venom of Crotali; but a single needle pricked through the mucous surface covered by the poison, sufficed to let in death.

It seems possible that minute ulcers or abrasions, quite invisible to the eye, might, in like manner, enable the venom in some cases to pass the barrier of the intestinal mucous lining. The deaths from ingested Cobra venom related by Fayrer took place in mammals, and we ourselves found in like experiments with rabbits, that, although death was rare after swallowing Cobra venom, it was less so than in pigeons; but our Cobra experiments are not strictly comparable with those done in India with fresh poison.

Certainly Cobra venom is much more apt to kill when swallowed than is Crotalus poison. In the rattlesnake it is the globulins which are in largest amount, and which are not dialysable, but in Cobra the fatal peptone is the material which, both as to vigor and amount, represents the poisoning capacity, and is as we know dialysable. It is only astonishing, therefore, that it does not kill in every case in which it is swallowed; but, as we have seen, the gastric juices in so far as they have time to act are destructive of venoms, and hence their protective agency has also to be considered.

The Activity of Venom when applied to Serous Surfaces.—One of the most remarkable and interesting of the physiological effects produced by the venom of the Crotalus is the occurrence of ecchymoses, especially in the serous tissues. The character of these ecchymoses is fully treated in another part of the work, so that we need here only detail some of our observations in connection with the direct effect of the application of venom to the serous tissues.

A rabbit was etherized and kept in this condition during the whole of the experiment. The abdominal cavity was opened and a knuckle of intestine exposed. On the peritoneum were placed a few small particles of the dried venom of the *Crotalus adamanteus*. In two or three minutes some extravasations appeared immediately about the point of the application of the venom; a few moments later these extravasations were diffused over a considerable area and had run into each other to such an extent as to form a patch of bleeding surface. So rapidly do these hemorrhages spread that they can literally be seen to grow under the eye. Another portion of the intestine was exposed, and upon the peritoneum was placed a very small portion of the glycerine solution of venom (prepared in 1862), which has already been referred to. Ten minutes after the application small points of extravasation appeared, and in three minutes more had increased so much in number and spread so rapidly as to form a continuous area of bleeding surface. Four drops of the glycerine solution of venom in a little water were boiled and carefully evaporated to a thick paste and then applied to a fresh surface of the peritoneum. After one hour no ecchymoses appeared.

In another experiment 0.03 gram of the dried Moccasin venom was boiled and injected into the peritoneal cavity of a pigeon. The animal died in forty-two minutes, when we found large ecchymoses scattered over all the abdominal viscera. In later experiments we have fully determined that the venom peptone may cause ecchymoses, but that this power exists in an insignificant degree as compared with that of the globulins.

In another experiment, not irrelevant here, we injected one drop of the fresh venom from the *Crotalus adamanteus* into one of the mesenteric arteries of an etherized rabbit. In a few seconds ecchymotic patches appeared on the large intestine followed by a few on the small intestine, and in another moment the animal was dead.

In an experiment on an alligator, elsewhere quoted, the activity with which venom may be absorbed by serous membranes is well illustrated. In a frog death occurred within two hours after the injection of two drops of the fresh venom of the *Crotalus adamanteus* into the peritoneal cavity.

In one experiment made upon an etherized rabbit in which 1 drop of fresh Moccasin venom was dissolved in 1.5 c. c. of distilled water and injected into the peritoneal cavity the animal died in one and a quarter hours. In an autopsy one hour after death, it was found that there was no rigor mortis; the whole of the inside of the peritoneal cavity was stained, and in places was literally dripping with blood; the mesentery contained a large amount of blood resembling a free clot. On the surface of the intestines the effusion of blood was of a brilliant red color as though from the arterioles; the whole interior of the abdominal cavity was stained; the heart was arrested in systole.

In still another experiment in which the solution of venom was boiled the results were strikingly different. We dissolved 0.03 gram of dried Moccasin venom (representing a much greater dose than was given in the previous experiment) and after boiling it for a moment filtered it. The filtrate was injected into the peritoneal cavity of a rabbit. The animal was killed after the lapse of one hour and the peritoneal cavity examined. There were absolutely no alterations to be seen in the viscera, excepting one minute spot where there appeared a little reddening.

The length of time during which the venom used was boiled was not distinctly stated in the notes of some of our observations. The omission was of moment.

At the time these experiments were made, we did not fully know that while in all venoms—brief boiling throws down the globulins at once—much longer boiling by degrees precipitates, and at last makes innocent the peptones. Apparently it is the globulins which most rapidly alter blood and vessels, and by a mechanism hereinafter to be described cause ecchymoses. Yet are the peptones not without this toxic capacity, as is seen in some of the above observations. Clearly, however, boiling impairs the activity of Crotaline peptones, as it does that of like constituents of Cobra poison. It will have been seen that none of these direct experiments on serous tissues were made with pure or boiled Cobra venom. It is desirable that this should be done, and especially with fresh venom. In another portion of this paper there are some relative studies of the power of dried Cobra and Rattlesnake venoms to cause local hemorrhages from the peritoneum. In the former work of Dr. Mitchell, and in that of Fayrer and Brunton, are sufficient studies of the absorbing power of rectal and pulmonary surfaces and of the eye.

CHAPTER V.

THE EFFECTS OF VENOM ON THE NERVOUS SYSTEM.

EXCEPTING as regards the marked action on the respiratory centres we cannot consider venom as essentially or solely a nerve poison. In animals which do not immediately die from the effects of this poison, the first signs of nerve poisoning are drowsiness, incoördination, followed by loss of voluntary motion, by convulsions, or failure of reflex activity and by death.

Reflex Action.—In six experiments on frogs with the Crotalus, made in connection with a direct study of the effect on reflex action, in none of them was there found a slow, gradual diminution of reflex activity, but invariably a sudden loss of this function. The time of the occurrence of the loss of reflex activity varies very greatly. In four experiments on pithed frogs, each of the frogs was given 0.015 gram of the dried *Crotalus adamanteus* venom in 10 minims of distilled water, by means of injection into the posterior lymph sac. In one experiment no alteration in reflex activity occurred after one and three-quarter hours, although it seems probable that the venom was not by any means completely absorbed since the lymph sac seemed bulged with fluid which had accumulated. In another experiment no alteration occurred in one and a half hours. In a third reflex action was suddenly abolished in one hour, and in a fourth in forty-five minutes, without there being in any case gradual diminution of reflex activity preceding the complete loss.

Two experiments were made on pithed frogs to determine if the loss of reflex activity was due to an action of the venom upon the nerves or upon the spinal cord, and for this purpose we ligated all of the bloodvessels in the right hind leg of each animal, and thus prevented the access of the venom to these parts. To each of these frogs was given 0.015 gram of the dried venom of the *Crotalus adamanteus* dissolved in 5 minims of distilled water, and injected into the posterior lymph sac. Reflex activity suddenly ceased in both of the frogs in one and a half hours. No reflex action was elicited by irritation of the nerves of either leg, although the motor fibres of the nerves were very excitable. We also found that direct excitation of the spinal cord in the dorsal region produced movements in the posterior extremities, but none in the anterior extremities, thus showing that impulses could travel down the cord through the motor apparatus but not upwards through the sensory portions. These observations make it clear that the loss of reflex activity is, no doubt, dependent to a great extent at least upon an action of the venom upon the sensory portions of the cord, although it is not clear that the sensory nerve fibres may not also be seriously affected.

Sensory and Motor Nerves.—In order to more directly test the action of venom upon the motor and sensory fibres we exposed the sciatic nerve along the whole extent of the thigh of pithed frogs, and placed in the middle of the exposed trunk a little (Crotalus) venom (concentrated by spontaneous evaporation). Comparative observations were now made from time to time by exciting the mixed nerve trunk above and below the point of application of the venom by means of electrodes connected with a Du Bois-Reymond induction coil, using minimum strengths of current. After about fifteen minutes, irritation of the foot of the leg with the poisoned nerve did not give as good reflexes as irritation of the other leg. After five hours, irritation of the trunk of the nerve below the poisoned part did not give reflexes, but above the part did give reflexes, showing that the sensory fibres were functionally destroyed by the local application of the poison. When the trunk was irritated above the poisoned part marked contraction of the muscles of the limb occurred, showing that the motor fibres and muscles were still intact. After the lapse of six hours the motor nerves would no longer respond to stimulus, although the muscles were still irritable.

From these observations it seems obvious that both the sensory and motor nerves are affected by the poison, and that the sensory nerves are far more susceptible than the motor nerves, and that the depression of the sensory nerves may be connected with the depression of reflex activity; but it seems more than likely that the loss of reflex activity is essentially of spinal origin, since there is not a slow, gradual diminution of reflex activity but a sudden paralysis—a characteristic which may be considered almost exclusively spinal.

The Spinal Cord.—We have already stated that the motor columns of the cord remain irritable after complete paralysis of the sensory columns. We have supplemented these observations by some experiments showing the direct action of venom upon the exposed spinal cord, which prove that the motor columns themselves ultimately succumb to the poison. In two experiments made upon large frogs, in which was laid bare the spinal cord in the dorsal region and in which the animals were left to fully recover from the shock, a concentrated solution of the dried venom of the *Crotalus adamanteus* was placed on a small portion of the cord. Before the application of the venom the cord responded actively to slight mechanical irritation; after the application of the venom there occurred a gradual impairment in irritability for the first fifteen minutes; this impairment increased, so that at the end of two hours the cord would not respond to moderate electrical stimulus. The diminution of function continued until at the end of seven hours the strongest current induced no response, although the motor nerve trunks responded actively.

Voluntary Motion.—Usually the earliest signs of nerve poisoning with venom are a disturbance of coördination and loss of voluntary motion. In frogs we found that as long as voluntary motion lasted the reflexes were active, but that with a loss of volition reflexes were at once decidedly diminished and suddenly disappeared. In frogs in which the abdominal aorta was ligated so as to prevent the poison from affecting the nerves of the posterior extremities, the results were similar.

In a number of observations made upon mammals the above conclusions were borne out. We have also found that the orbital reflexes are completely gone before death, and that before their loss voluntary motion disappears. Moreover, if the spinal cord is irritated immediately after the cessation of the orbital reflexes it will be found that irritation will give rise to movements in the posterior extremities only, and that after the cord will no longer respond to irritation the motor nerves are still excitable. After the motor nerves cease to respond the muscles remain irritable.

These results all go to establish the conclusion that the respiratory centre is the most vulnerable point of the nervous system, that the coördinating and volitional centres are then prominently affected, that the sensory part of the spinal cord and the sensory nerves are next attacked, and that the motor parts of the cord, and the motor nerves are the last to succumb.

CHAPTER VI.

THE GLOBULINS AND PEPTONES COMPARED AS REGARDS LOCAL POISONOUS ACTIVITY.

It seems needful at this place to consider the relative local toxic capacity of globulins and peptones, the two substances found in varying quantities in all venoms as yet examined.

In order to do this effectively it will be needful at the risk of anticipating a part of what belongs strictly speaking to the pathological section, to speak briefly of the macroscopical lesions brought about at the seat of injection by these potent substances.

What takes place intensely where the injection needle enters, but represents in a violent and coarser manner lesions to be found soon or late throughout the body, and this especially applies to entire venom and to the globulins. These studies of local changes are not without definite explanatory value. It has long since been shown that the cobra and the rattlesnake are distinctive poisoners, and now our latest work seems to explain just why this is so, and enables us to see already that what might efficiently aid one bitten by the Indian serpent, would be by no means sure to succor the victim of our own less fatal snake.

Venom Peptones. Local Action.—The albuminous elements of venoms are, as already shown, two in number, and belong by virtue of their reactions respectively to the classes, peptone and globulin. Hence, as we now see with clearness, it is easy to separate them by boiling, *which if brief,* destroys the globulin as a poison and leaves the peptone unaltered. When after boiling we inject the fluid and coagula, we still poison, if the dose be large, for the venom peptone is toxically unchanged. The wound shows, however, hardly any of the singular appearances which characterize lesions due to fresh or unboiled venom. Boiling leaves the poison less active locally. *If continued* it also affects more or less the general toxicity, but this influence is most marked in the Crotaline venoms, because in them the peptone is least in amount and is also the least deadly of the two constituent poisons.

Venom Peptone.—When venom peptone in full dose is injected into the breast of a pigeon, if the animal dies within an hour or two, there is scarcely any appreciable local effect, as will be clearly seen by examining the results of experiments recorded in the chapter on the influence of various agents on the poisonous activity of venom. If the dose be smaller, so that life is prolonged, the first local effect observed is a considerable œdematous swelling without any dark discoloration. After the lapse of about eighteen hours there is apt to be some discoloration, and generally a discharge of muddy putrescent serum. If the animal be killed after a day it will be found that the muscles on the injected side about the region of the œdema are pale and bloodless, having the appearance of half-cooked chicken meat. In animals which lived longer there was sometimes found considerable congestion, marked by greenish streaks, and giving off horrible putrefactive odors. In others beneath the œdematous swelling lay a cavity about an inch in diameter, which was full of broken-down tissue, having a muddy, gangrenous appearance, and decidedly putrescent, while the surrounding muscular tissues were not apparently altered in appearance. In none of these experiments were ecchymoses found in the intestines, and in all of them the blood was coagulable.

In the following experiment Crotalus peptone obtained by dialysis was at 2:37 injected into the breast of a pigeon; 3:15 weak; 3:50 rocking slightly, no local discoloration, some slight œdematous swelling; 4:00 more unsteady on its feet; 5:45 there was considerable watery effusion in the subcutaneous cellular tissue on the side of the injection. The following afternoon there was a large swelling over the site of the wound. It was an inch or more above the healthy skin, and was apparently purely œdematous in character, there being no dark discoloration nor appearances of congestion. The superficial local effect was in every way unlike that produced by the globulins. The following afternoon (after 48 hours) the pigeon died. The swelling was unaltered as to size, and but very little discolored. The tissues around it were slightly darkened, the coloration fading away gradually at about one-half inch from the border, and there was a well-defined pale streak of tissue between the swelling and the surrounding tissue like a line of demarcation. Upon cutting into the tumor serum dropped from the incision, and the subcutaneous cellular tissue was found greatly infiltrated. The swelling seemed to be almost entirely œdematous and the serum had a putrefactive odor. The muscular tissues were greatly congested and somewhat blackened, and in places as green as though infiltrated with bile. This green appearance could be seen distinctly through the skin on the surface of the superficial muscles, extending over the entire side of the breast. The odor emanating from the cut muscles was also putrefactive. In the intermuscular tissues there was some greenish gelatinous matter. Beneath the swelling was a streak of muscular tissue about one-fourth of an inch thick, which was very pale, like half-stewed meat, contrasting strongly with the other parts of the muscles.

All of these observations on slow poisoning were made with peptone derived from the venoms of the *Crotalus adamanteus* or the Moccasin.

Judging from the fact that the venom peptone does not give rise to any darkening of the muscular tissues within a short time after injection, and indeed, as it seems probable, not until putrefaction has set in, it is likely that the darkening and congestion which ultimately occur are to be regarded as mere secondary effects, and due to putrefactive changes induced by the poison.

The peptones, whether obtained by boiling or dialysis, seem to cause locally an enormous œdema, gradual breaking down of the tissues, and rapid production of horrible putrefactive processes, with finally a more or less extensive slough. They possess little power to produce large hemorrhages, because they do not so well as venom globulin destroy the coagulability of the blood. Hence in peptone wounds there are only such local bleedings as are due to the leakage caused by gangrenous processes. The ragged, sodden grayish look of the muscles is very remarkable, and once seen is too unfamiliar not to be remembered as a most striking pathological appearance. For effects of peptone, see Plate I.

Venom Globulins. Local Influence.—When we inject unboiled venom, we are using globulin as well as peptone, in amounts which differ with every serpent. If we use the isolated globulins the contrast in the local phenomena as compared with those caused by peptones is immense. The different globulins already described were all examined in this connection.

As the globulins are insoluble in water free from salines, dialysis kept up long enough, as from forty-eight to seventy-two hours, in a temperature so low as to insure absence of putrefaction, will throw down the mass of the globulins in a form which enables us to collect and re-dissolve them. Three drops of Moccasin venom were mixed with 6 c. c. distilled water and dialysed by a current of pure water for fifty-six hours. As the salts passed out a precipitate increased within the dialyser. After having been washed with distilled water, it was thrown into the breast of a pigeon. Death took place in twenty-four hours. This delay in a fatal result was owing to the dose being small, and perhaps also to the fact that it did not represent all the globulin of three drops of venom; after death there was a tense black swelling at the site of the wound, and the tissues, for two inches in every direction, were soaked with black absolutely fluid blood.

It is difficult to subject venom constituents to any processes like solution or drying without more or less altering their toxicity. Desiccation certainly affects whole venom, and in a measure lessens the severity of its local symptoms. The same is true of venom globulins. An equal dose of globulin dried and redissolved takes longer to kill, than if not previously dried; also if the dried venom be given in unusual dose, the local effects are slighter than those seen with pure venom or fresh globulin in smaller dose, but killing within the same time. A long survival of course enables the local phenomena to develop and might mislead as to the fact of drying having an enfeebling influence. Desiccation greatly lessens the solubility of venom, and of its albuminous constituents, and in consequence they fail to permeate the tissues and to enter the blood at the rate which characterizes fresh venom.

In the experiment which follows, death was long delayed, and owing to this the local results were strongly marked.

A quantity of globulin obtained by dialysis, representing two grains of the venom of the *Crotalus adamanteus*, was allowed to dry. It was then placed in a little distilled water, and after a few minutes a small amount of common salt was added, which caused the venom and water to form a milky solution.

This was injected into the breast muscles of a pigeon at 3:25; 3:40 some darkening and swelling of the side of injection; 4:25 unable to stand; 6:00 convulsions, followed by death.

Autopsy.—The local effect of the venom was remarkable; beneath the skin in the areolar tissue, over the wounded side and over half the breast of the opposite side, was a mass of bloody gelatinous effusion, and the muscles beneath on the injected side were swollen and darkened and enormously infiltrated with blood. See Plate II., Fig. 1.

These experiments, which have been supplemented by many others, give a somewhat definite idea of the marked difference in the local effects of the globulins as a group in comparison with those produced by the boiled solution of venom, or in other words by the venom peptone.

Experiment.—The *water-venom-globulin* from 0.03 gram of dried venom of the *Crotalus adamanteus*, dissolved in a little water by means of a few crystals of salt, was injected into the breast muscles of a pigeon at 3:55. At 5:50 the animal was dead. The region of injection was terribly swollen, blackened, and suffused with liquid blood. (The amount of globulin injected was about 0.003 gram.)

In a rabbit to which had been given some of the *water-venom-globulin* intravenously, enormous extravasations were found when the abdominal cavity was opened.

In another experiment with the *water-venom-globulin* obtained from the venom of the Moccasin, the animal lived for some time, and very characteristic effects of slow poisoning from venom globulin were observed.

Experiment.—One c. c. of distilled water containing 0.001 gram of *water-venom-globulin* from the Moccasin was thrown into the breast muscles of a pigeon at 5:26. At 6:00 the local darkening and swelling of the tissues at the region of injection were noticeable. After twenty-four hours the animal was in a generally fair condition; the side was considerably darkened, and on the breast was a large swelling, which appeared to be due to a bloody effusion into the subcutaneous tissue. After forty-eight hours there was a discharge of red serum with a putrefactive odor. The whole of the side was darkened and greenish, and had the appearance of commencing gangrene.

Copper-venom-globulin.—Some of the *copper-venom-globulin* from the venom of the *Crotalus adamanteus* was injected with a little water into the breast muscles of a pigeon at 4:35. At 5:00 it was weak, but no local effects were apparent. On the following morning it was dead. The local effects were intense; there was considerable swelling, blackening, and diffusion of fluid blood. The heart was arrested midway between systole and diastole, and contained fluid blood of a dark color.

Dialysis-venom-globulin.—Some of the *dialysis-venom-globulin* from the venom of the *Crotalus adamanteus* was injected into the breast muscles of a pigeon at 5:03. At 5:18 was sickish; 5:19 unsteady, side somewhat swollen and darkened; 5:30 local effect increased. Following morning—dead. Local effects intense—great swelling, blackening, and diffusion of blood, which is incoagulable.

In another experiment, in which a larger quantity was used, the bird died in twenty-five minutes, after the occurrence of stupor, incoördination, deep laborious breathing, and convulsions. There was no time for very decided local effects, but the blood was tarry and incoagulable.

These experiments, which have been frequently repeated, render it clear that the remarkable local effects produced by the venoms of the Crotalus and Moccasin, and which are not observed after the venom is boiled, are due to the venom globulins, all of which bring about essentially the same local alterations. To fully satisfy ourselves that these interesting local effects were dependent upon the physiological activities of the globulins and not upon a possible contamination by peptone, observations were made with the *boiled* globulins. In none were there the least evidences of the presence of any poisonous element.

Whenever globulins alone are used, we have these local bleedings, fluid blood, and capillaries giving way soon after the poison reaches them. The system at large soon or late repeats the coarser phenomena of the wound; and yielding vessel walls, fluid blood, and countless hemorrhagic outflows exhibit the power of the globulins. Peptone, or, which is much the same, briefly-boiled venom, causes putrefactive changes swiftly, and shows but slight capacity to make fluid the blood, or to corrode the capillaries. The wound is foul and œdematous, but not filled with blood, whilst in its general effects the venom peptone fails again to exhibit the capacity of the globulins to multiply hemorrhages, and to destroy the natural ability of the blood by clotting to check its own wasteful expenditure.

In proportion as the peptones predominate will we have then a lessening of rapidly formed local lesions, and this is of course why Cobra venom does not give us the same terrible local consequences which ensue in *Daboia*, *Moccasin*, and *Crotalus* bites, where we have the potent combination of enough peptones and an excess of globulins. For a comparison of the local effects of Cobra and Crotalus poisoning, see Plate II., Figs. 2 and 3.

CHAPTER VII.

THE ACTION OF VENOMS AND THEIR ISOLATED GLOBULINS AND PEPTONES UPON THE PULSE-RATE.

SECTION I.—PURE VENOM.

THE experiments made in connection with the pulse-rate were performed upon rabbits, and in every case, unless otherwise noted, the poison was dissolved in 1 c. c. of distilled water and injected intravenously, usually into the external jugular vein.

In researches made with the isolated poisons doses were usually employed which represented the amount of the individual poison contained in the commonly employed doses of the pure dried venom, thus giving a fair idea of the part played by the individual principles in the results produced. In some experiments, however, much larger doses were used to learn more fully the poisonous character of these substances.

In all of our observations we find that the results produced in animals, under apparently the same conditions and by using the same doses, vary very greatly; sometimes the pulse is quickened from the first and remains beyond the normal until death ensues, sometimes there is a primary diminution followed by an increase, at others there is a diminution which continues until death. The pulse is generally found to vary much in frequency. These facts all suggest that the action of the pure venom is of a complex nature; there being several factors concerned in the various alterations, and render it not improbable that in some instances ecchymoses in the various organs may account for exceptional variations.

Twenty experiments were made with pure venoms upon normal animals; six of these were made with the venom of the *Crotalus adamanteus*; in three the pulse-rate was diminished and remained below normal, in two there was a primary increase followed by a diminution, and in one of these the pulse-rate afterwards went above the normal, while in another there was a primary diminution followed by an increase. Of two experiments made with the *Crotalus horridus*, in one there was an increase which continued until death, and in the other an increase followed by a diminution below the normal, this diminution in turn being followed by a rise above the normal, which continued until approaching death. In two experiments with the *Ancistrodon piscivorus*, in one there was an increase and in the other a decrease. One experiment with the *Ancistrodon contortrix* gave an increase. In one experiment with the *Crotalophorus miliaris* there was a decrease followed by an increase. In one with the *Daboia Russellii*, which was not a perfectly satisfactory experiment, there was a decrease, and in six experiments with the Cobra there was an increase in all, the increase being followed in three by a permanent decrease; in one the increase was followed by a diminution, and this in turn by an increase; in two experiments there was a permanent increase, excepting near death when a decrease ensued.

It will thus be clear that even under apparently the same conditions we cannot foretell what the alterations in the pulse-rate will be in any given experiment; although the results of the six experiments with Cobra venom are so uniform in regard to the primary increase as to indicate that with it at least we should always expect to find more or less acceleration which may or may not continue above normal, even up to the time of death.

We may also add here, that we cannot trace any relations in the alterations in the pulse, arterial pressure, and respiration to each other, so that it seems as if the changes must depend essentially upon actions peculiar to each apparatus. This holds good with the study of the pure venoms or their isolated poisons.

Action of the Pure Venoms upon the Pulse-rate in Normal Animals.

Experiment No. 1.

	Time: min. sec.	Pulsations per minute.	REMARKS.
Normal	. . .	285	Injected 1 drop of fresh venom from the *Crotalus adamanteus* into the thigh of a large rabbit.
	20	285	
	40	285	
	1 00	285	
	1 20	285	
	1 40	285	
	2 00	?	Clot in canula.
	5 00	285	
	7 00	285	
	8 00	285	
	9 00	270	
	10 00	270	
	11 00	240	
	12 00	255	
	13 00	270	
	20 00	270	
	21 30	270	
	23 00	270	At 1:00 the blood-pressure began falling and reached a minimum at 10:00, when it was one-third less than the normal.

Experiment No. 2.

	Time: min. sec.	Pulsations per minute.	REMARKS.
Normal	. . .	240	Injected 3 drops of the fresh venom of the *Crotalus adamanteus* into the thigh of a large rabbit.
	20	240	
	40	240	
	1 00	195	
	1 20	. . .	Animal broke loose and **tore** the canula from the **artery**.
8 April, 1886.			

Experiment No. 3.

	Time: min. sec.	Pulsations per minute.	REMARKS.
Normal	. . .	255	Injected intravenously 0.003 gram dried venom of the *Crotalus adamanteus* in 1 c. c. distilled water.
	10	240	
	20	225	
	30	225	
	40	210	
	50	195	
1	00	195	
1	20	210	
1	40	210	
2	00	225	
2	20	270	
2	40	270	

Experiment No. 4.

	Time: min. sec.	Pulsations per minute.	REMARKS.
Normal	. . .	300	Injected intravenously 0.003 gram dried venom of the *Crotalus adamanteus* dissolved in 1 c. c. distilled water.
	5	310	
	10	330	
	20	330	
	30	300	
	40	270	
	50	270	
1	00	270	
1	20	280	
1	40	270	
2	10	270	
4	00	280	
7	00	300	
8	00	90	Struggles accompanied by remarkably slow heart beats and considerable increase of arterial pressure.
8	10	130	
8	20	170	
8	30	240	
8	40	292	
8	50	320	
10	30	320	
13	00	280	
13	30	. . .	Injected 0.003 gram dried venom dissolved in 1 c. c. distilled water.
13	50	300	
14	00	300	
16	00	280	
16	05	. . .	Repeated the injection.
16	30	280	
17	00	. . .	Dead. Heart in complete diastole; blood incoagulable, ecchymoses in pericardium and peritoneum.

THE ACTION OF VENOMS UPON THE PULSE-RATE.

Experiment No. 5.

Time: min. sec.	Pulsations per minute.	REMARKS.
Normal . . .	225	Injected intravenously 0.015 gram dried venom of the *Crotalus adamanteus* dissolved in 1 c. c. distilled water.
10	270	
20	120	} Struggles.
30	180	
40	285	
1 00	285	
1 20	285	
1 40	285	
2 00	285	
4 00	?	Too feeble to count.
8 00	..	Dead.

Experiment No. 6.

Time: min. sec.	Pulsations per minute.	REMARKS.
Normal . . .	323	Injected intravenously 0.015 gram dried venom of the *Crotalus adamanteus* dissolved in 1 c. c. distilled water.
5	315	
20	214	
30	225	
40	255	
1 00	255	
1 30	255	
3 30	240	
5 30	240	
7 30	225	
8 00	195	
13 00	195	
17 30	. . .	Dead.

Experiment No. 7.

Time: min. sec.	Pulsations per minute.	REMARKS.
Normal . . .	260	Injected intravenously 0.02 gram dried venom of the *Crotalus adamanteus* dissolved in 1 c. c. distilled water.
5	260	
20	98	Blood pressure increased, probably due to asphyxia.
30	84	
40	120	
1 00	. . .	Dead.

Experiment No. 8.

Time: min. sec.	Pulsations per minute.	REMARKS.
Normal . . .	300	Injected into the *right carotid artery* 0.015 gram dried venom of the *Crotalus adamanteus* dissolved in 1 c. c. distilled water.
10	195	
30	60	
44	70	
55	130	
60	220	
80	260	
1 40	. . .	Dead.

Experiment No. 9.

Time: min. sec.		Pulsations per minute.	REMARKS.
Normal	...	225	Injected intravenously 0.015 gram dried venom of the *Crotalus horridus* dissolved in 1 c. c. distilled water.
	10	235	
	20	240	
	30	240	
	50	240	
1	00	240	
1	10	240	
1	20	240	
1	30	240	
1	40	240	
3	40	260	
5	40	280	
7	40	330	
9	40	350	Convulsions.
10	10	...	Dead.

Experiment No. 10.

Time: min. sec.		Pulsations per minute	REMARKS.
Normal	...	270	Injected intravenously 0.015 gram dried venom of the *Crotalus horridus* dissolved in 1 c. c. distilled water.
	5	...	
	10	...	
	20	...	
	30	...	Animal broke loose and was replaced—record before this time valueless.
2	40	370	
3	00	400	
5	00	350	
6	30	225	
7	30	240	
8	00	280	
8	30	280	
9	00	330	
10	00	90	Respirations cease.

Experiment No. 11.

Time: min. sec.		Pulsations per minute.	REMARKS.
Normal	...	216	Injected intravenously 0.004 gram dried venom of the *Ancistrodon piscicorus* dissolved in 1 c. c. distilled water.
	20	228	
	30	228	Convulsions.
	40	245	
1	00	252	
1	30	240	Injected a similar dose.
1	50	284	
2	10	280	
2	30	280	
3	00	280	
4	00		Killed by pithing.

Experiment No. 12.

	Time: min. sec.	Pulsations per minute.	REMARKS.
Normal	. . .	305	Injected intravenously 0.004 gram dried venom of the *Ancistrodon piscivorus* dissolved in 1 c. c. distilled water.
	10	300	
	20	240	
	30	240	
	1 00	270	
	1 30	265	
	2 00	265	
	2 30	270	
	3 00	275	
	5 00	300	
	5 30	300	Injected as in the foregoing.
	5 35	300	
	5 45	300	
	6 05	260	
	6 15	. . .	} Convulsive movements
	6 25	. . .	
	6 45	250	
	7 25	145	
	7 35	130	
	7 45	120	
	7 55	100	
	8 05	115	
	8 15	120	
	8 25	132	Animal died in a few minutes.

Experiment No. 13.

	Time: min. sec.	Pulsations per minute.	REMARKS.
Normal	. . .	320	Injected intravenously 0.003 gram dried venom of the *Ancistrodon contortrix* dissolved in 1 c. c. distilled water.
	10	320	
	30	370	
	1 00	340	
	1 30	340	
	2 00	330	
	2 30	330	
	4 30	360	
	5 30	350	Injected as in the foregoing.
	5 50	320	Struggles.
	6 30	250	
	6 50	360	
	7 50	390	Injection repeated.
	10 00	. . .	Death.

Experiment No. 14.

Time: min. sec.	Pulsations per minute.	REMARKS.
Normal . . .	280	Injected intravenously 0.003 gram dried venom of the *Crotalophorus miliaris* dissolved in 1 c. c. distilled water.
20	285	
30	128	
40	135	
50	240	**Struggles.**
1 00	285	
1 10	285	
2 10	290	
2 40	300	Injected 0.006 gram.
2 50	300	
3 20	250	
5 20	275	
7 20	300	
9 20	315	

Experiment No. 15.

Time: min. sec.	Pulsations per minute.	REMARKS.
Normal . . .	240	Injected intravenously 0.003 gram dried Cobra venom dissolved in 1 c. c. distilled water.
10	240	
30	250	
1 00	260	
1 20	250	
3 20	250	
5 20	260	

Experiment No. 16.

Time: min. sec.	Pulsations per minute.	REMARKS.
Normal . . .	205	Injected intravenously 0.003 gram dried Cobra venom dissolved in 1 c. c. distilled water with a few crystals of sodic chloride.
1 00	205	
3 00	203	
8 00	205	
10 00	126	
15 00	165	
17 00	105	
19 00		Dead.

Experiment No. 17.

Time: min. sec.	Pulsations per minute.	REMARKS.
Normal . . .	260	Injected intravenously 0.005 gram dried Cobra venom dissolved in 1 c. c. distilled water with a few crystals of sodic chloride.
20	270	
30	250	
40	250	
1 00	250	
1 20	250	
1 40	230	
4 40	240	
7 40	250	
8 40	190	
9 10	. . .	Clot in canula.
15 00	. . .	Dead.

THE ACTION OF VENOMS UPON THE PULSE-RATE.

Experiment No. 18.

Time: min. sec.	Pulsations per minute.	REMARKS.
Normal . . .	310	Injected intravenously 0.015 gram dried Cobra venom dissolved in 1 c. c. distilled water.
10	310	
20	320	
30	310	
40	260	
1 00	310	
1 30	330	
2 00	340	
2 10	340	
6 20	150	
6 50	150	
7 30	165	
8 20	. . .	Dead.

Experiment No. 19.

Time: min. sec.	Pulsations per minute.	REMARKS.
Normal . . .	290	Injected intravenously 0.005 gram dried venom of the *Daboia Russellii* dissolved in 0.5 c. c. distilled water.
10	290	
15	280	
20	280	
30	280	
40	. . .	Tetanic convulsions.
2 00	. . .	Dead.

Experiment No. 20.

Time: min. sec.	Pulsations per minute.	REMARKS.
Normal . . .	216	Injected intravenously 0.003 gram dried venom of the *Cobra* dissolved in 1 c. c. distilled water.
0	. . .	
13	. . .	
20	252	
40	252	
1 00	264	
1 30	288	
2 00	260	
4 00	264	
9 00	231	
12 00	108	
14 00	48	Respiration ceased; artificial respiration used.
20	126	

Actions of Pure Venoms on the Pulse-rate in Animals with Cut Pneumogastric Nerves.—After section of the pneumogastric nerves we invariably found an increase which was, as a rule, very slight. Seven experiments in all were made on animals thus operated upon: one with the venom of the *Crotalus adamanteus;* one with the *Crotalus horridus;* one with the *Ancistrodon piscivorus;* one with the *Ancistrodon contortrix,* and three with the *Cobra.*

Experiment No. 21.

Time: min. sec.	Pulsations per minute.	REMARKS.
Normal . . .	295	Pneumogastric nerves cut. Injected intravenously 0.003 gram dried venom of the *Crotalus adamanteus* dissolved in 1 c. c. distilled water.
10	300	
30	300	
1 00	300	
1 30	295	
2 00	300	
2 30	305	
5 30	290	
7 00	285	
8 30	. . .	Injected a similar dose.
9 30	. . .	Dead.

Experiment No. 22.

Time: min. sec.	Pulsations per minute.	REMARKS.
Normal . . .	315	Pneumogastric nerves cut. Injected intravenously 0.015 gram dried venom of the *Crotalus horridus* dissolved in 1 c. c. distilled water.
20	. . .	Violent struggles.
1 00	320	
1 20	335	
1 40	330	
2 00	320	
2 20	340	
4 20	340	
5 50	350	
8		Dead.

Experiment No. 23.

Time: min. sec.	Pulsations per minute.	REMARKS.
Normal . . .	240	Pneumogastric nerves cut. Injected intravenously 0.003 gram dried venom of the *Ancistrodon piscivorus* dissolved in 1 c. c. distilled water.
20	300	
30	300	Struggles.
40	300	
50	300	
1 00	240	
1 20	225	
3 50	210	Struggles.
5 50	270	
6 00	300	Convulsions.
8 00	285	
10 00	270	
12 00	270	
14 00	285	
19 00	240	Injected a similar dose.
19 10	255	
19 20	255	
19 30	255	
19 40	240	
19 50	240	
20 00	240	

THE ACTION OF VENOMS UPON THE PULSE-RATE.

Time: min. sec.	Pulsations per minute.	REMARKS.
20 20	240	
21 20	255	
23 20	300	
25 20	270	
28 20	255	
33 20	255	
38 20	255	
44 20	255	Injected a similar dose.
44 40	225	
45 10	...	Dead.

Experiment No. 24.

	Time: min. sec.	Pulsations per minute.	REMARKS.
Normal	...	300	Pneumogastric nerves cut. Injected intravenously 0.003 gram dried venom of the *Ancistrodon contortrix* dissolved in 1 c. c. distilled water.
	10	300	
	20	310	
	30	310	
	40	310	
	1 00	310	
	1 50	310	
	4 20	310	
	7 00	310	Injected a similar dose.
	7 05	...	Struggles.
	7 10	320	
	7 20	300	
	7 40	300	
	8 00	310	
	9 00	315	
	11 30	300	Injected a similar dose.
	11 40	295	
	12 00	290	
	12 30	290	
	13 00	290	
	13 30	290	
	19 00	285	
	19 20	270	
	19 40	270	
	20 20	285	
	21 50	285	
	22 50	285	

Experiment No. 25.

	Time: min. sec.	Pulsations per minute.	REMARKS.
Normal	...	330	Pneumogastric nerves cut. Injected intravenously 0.003 gram dried Cobra venom dissolved in 1 c. c. distilled water with a few crystals of sodic chloride and filtered.
	10	330	
	30	330	
	1 00	330	
	3 30	340	
	6 30	330	
	10 30	320	
	14 30	295	
	16 30	275	Clot formed in canula.

9 May, 1886.

Experiment No. 26.

Time: min. sec.	Pulsations per minute.	REMARKS.
Normal . . .	320	Pneumogastric nerves cut. Injected intravenously 0.006 gram dried Cobra venom prepared as in the foregoing experiment.
10	315	
30	330	
1 00	330	
1 30	330	
3 30	335	
5 30	340	
9 30	320	
11 30	330	
18 30	. . .	Convulsions; asphyxia; death in 2½ minutes.

Experiment No. 27.

Time: min. sec.	Pulsations per minute.	REMARKS.
Normal . . .	390	Pneumogastric nerves cut. Injected intravenously 0.003 gram dried Cobra venom dissolved in 1 c. c. distilled water.
0	. . .	
10	. . .	
20	396	
1 00	390	
2 00	360	
4 00	354	
10	. . .	Clot in canula.
15	. . .	Dead of asphyxia.

The Actions of Pure Venoms on Animals in which Sections of the Pneumogastric Nerves and of the Upper Cervical Portion of the Spinal Cord had been made.—After isolation of the heart from the nerve centres by making section of the pneumogastric nerves and spinal cord in the middle or upper cervical region, and maintaining the animal alive by means of artificial respiration, we find that the pulsations of the heart are almost invariably slightly diminished in frequency upon use of venom. Seven experiments were made: three with the venom of the *Crotalus adamanteus;* one with the *Crotalus horridus;* one with the *Ancistrodon piscivorus;* one with the *Ancistrodon contortrix,* and one with *Cobra.* In one experiment with the *Crotalus adamanteus* in which two doses were given, there occurred a diminution after the first dose, while there was a marked increase after the second. In the experiment with the *Crotalus horridus* there was but little alteration.

Experiment No. 28.

Time: min. sec.	Pulsations per minute.	REMARKS.
Normal . . .	240	Pneumogastric nerves and cord cut. Injected intravenously 0.003 gram dried venom of the *Crotalus adamanteus* dissolved in 1 c. c. distilled water.
10	235	
20	230	
30	225	
40	215	
1 00	210	
1 20	210	
1 40	210	
2 40	. . .	Dead.

THE ACTION OF VENOMS UPON THE PULSE-RATE.

Experiment No. 29.

Time: min. sec.	Pulsations per minute.	REMARKS.
Normal . . .	185	Pneumogastric nerves and cord cut. Injected intravenously 0.003 gram dried venom of the *Crotalus adamanteus* dissolved in 1 c. c. distilled water.
10	185	
20	185	
30	180	
1 00	180	
1 20	180	
3 20	180	
3 50	180	
5 50	160	
7 50	. . .	Dead.

Experiment No. 30.

Time: min. sec.	Pulsations per minute.	REMARKS.
Normal . . .	240	Pneumogastric nerves and cord cut. Injected intravenously 0.003 gram dried venom of the *Crotalus adamanteus* dissolved in 1 c. c. distilled water.
10	230	
20	230	
30	230	
40	195	
1 00	200	
1 20	205	
1 40	210	
2 00	230	
2 30	265	
3 00	250	
6 00	300?	Injected a similar dose.
6 05	265	
6 15	. . .	
6 35	270	
7 05	260	
7 35	260	
8 05	260	
15 05	. . .	Dead.

Experiment No. 31.

Time: min. sec.	Pulsations per minute.	REMARKS.
Normal . . .	235	Pneumogastric nerves and cord cut. Injected intravenously 0.015 gram dried venom of the *Crotalus horridus* dissolved in 1 c. c. distilled water.
10	230	
30	240	
40	240	
1 00	235	
2 00	240	
4 00	240	
6 00	230	
8 00	220	
12 00	. . .	Dead.

Experiment No. 32.

Time: min. sec.	Pulsations per minute.	REMARKS.
Normal . . .	220	Pneumogastric nerves and cord cut. Injected intravenously 0.003 gram dried venom of the *Ancistrodon piscivorus* dissolved in 1 c. c. distilled water.
20	210	
30	200	Struggles.
40	200	
1 00	210	
1 20	210	
1 50	210	
4 20	195	
8 20	210	
10 20	210	
12 20	210	
15 20	210	
18 20	210	
21 20	. . .	Dead.

Experiment No. 33.

Time: min. sec.	Pulsations per minute.	REMARKS.
Normal . . .	260	Pneumogastric nerves and cord cut. Injected intravenously 0.003 gram dried venom of the *Ancistrodon contortrix* dissolved in 1 c. c. distilled water.
10	255	
20	250	
40	243	
1 00	243	
1 30	245	
2 00	240	
4 00	240	
7 00	240	
9 00	240	
11 00	240	
13 00	240	
15 00	240	
17 00	240	
20 00	240	
22 00	240	Injected 0.006 gram.
22 15	?	
22 30	?	Dead.

Experiment No. 34.

Time: min. sec.	Pulsations per minute.	REMARKS.
Normal . . .	220	Pneumogastric nerves and cord cut. Injected intravenously 0.003 gram dried Cobra venom dissolved in 1 c. c. distilled water.
10	215	
30	215	
1 00	210	
3 00	215	
5 00	215	
8 00	225	
11 00	225	
14 00	225	
19 00	225	Killed by pithing.

Summary and Conclusions of the Actions of Venoms on the Pulse-rate.—The results of this series of experiments indicate that the primary tendency of venoms is to cause an increase of the pulse-rate, that this tendency is greater after section of the pneumogastric nerves, and that it rarely occurs after conjoined section of the pneumogastric nerves and the upper or middle cervical region of the spinal cord.

From the increased tendency to acceleration of the pulse-rate in poisoning by venom after section of the pneumogastric nerves we infer that there is some direct or indirect effect of the venom upon the pneumogastric centres by which an inhibitory influence is exerted, and which tends to neutralize the action bringing about acceleration. Since hastening of the pulse is a rare occurrence after conjoint section of the pneumogastric nerves and the cervical spinal cord, we think that the increase is due for the most part to some effect upon the accelerator centres in the medulla, whereby impulses are sent through (chiefly at least) those of the accelerator fibres which pass by the cord. The increase of the pulse-rate which may occur after division of the nerves distributed to the heart, by section of the pneumogastric nerves and cervical spinal cord, must be dependent upon a direct action of the venom upon the heart muscle or its contained ganglia.

The diminution in the heart beats must be due to a direct cardiac action, since it occurs after isolation of the heart, as above, from any central nervous influence.

In these as in all other experiments which involve intravenous use of venoms we are liable to disturbing elements which do not trouble our explanations in dealing with other poisons. At any moment, anywhere in nerve-tissue or muscles, we may have abrupt and quite countless hemorrhages. How these may introduce conflicting symptoms and modify results has already been pointed out by one of us many years ago.[1] They make absolute constancy of effects quite improbable.

Section II.—The Actions of Globulins on the Pulse-Rate.

The Actions of the Venom Globulins on the Pulse-rate.—The actions of the venom globulins upon the pulse-rate appear to differ somewhat in quality from what is found in poisoning with pure venoms; there is a greater tendency to the primary increase in the pulse than with pure venoms, while the action by which this is brought about seems to differ.

Of eleven experiments in which the amounts used represented the proportion of the respective globulins contained in the usual doses of venom given, six were made with the water-venom-globulin, two with the copper-venom-globulin, and three with the dialysis-venom-globulin; all of these poisons, excepting in one experiment with the water-venom-globulin of the Ancistrodon, were derived from the venom of the *Crotalus adamanteus*.

The water-venom-globulin seems to be the most active, and the copper-venom-

---

[1] Researches on the Venom of the Rattlesnake. S. Weir Mitchell, 1861.

globulin the least so. Of the six experiments with the former, in four there was a primary increase in the pulse-rate followed by diminution, and in one case by a subsequent increase; in the other two there was a diminution from the first, the pulse regaining its normal frequency, or, as in one instance, rising above it.

In both the experiments with copper-venom-globulin there was a primary increase followed by a diminution in one case, and in the other by a return of the rate to about the normal.

In the three experiments with dialysis-venom-globulin, a primary increase occurred. In two this was followed by a drop below normal, while in the other the rate remained above the normal.

Experiment No. 35.

	Time: min. sec.	Pulsations per minute.	REMARKS.
Normal	. .	290	Injected intravenously 0.0012 gram *water-venom-globulin*
	10	305	(= 0.015 gram dried venom) from the venom of the *Cro-*
	20	310	*talus adamanteus*.
	40	315	
	1 00	290	
	1 30	270	
	3 00	270	
	5 00	270	
	7 00	270	
	9 00	280	
	12 00	290	
	15 00	300	
	18 00	315	
	25 00	320	
	35 00	330	
	45 00	330	
	55 00	330	Killed.

Experiment No. 36.

	Time: min. sec.	Pulsations per minute.	REMARKS.
Normal	. . .	310	Injected intravenously the *water-venom-globulin* from 0.03
	10	310	gram dried venom of the *Crotalus adamanteus*.
	20	275	
	40	**265**	
	1 00	260	
	1 20	260	
	1 40	280	
	3 40	290	Clot in canula.
	7 40	310	
	9 40	310	Injected *water-venom-globulin* from 0.015 gram dried venom
	10 00	315	
	10 20	300	
	10 40	300	
	14 00	300	
	17 00	300	
	20 00	300	
	30 00	300	Killed by pithing.

Experiment No. 37.

Time: min. sec.		Pulsations per minute.	REMARKS.
Normal	. . .	320	Injected intravenously 0.0033 gram *water-venom-globulin*
	10	350	from the dried venom of the *Crotalus adamanteus* dissolved
	20	330	by the addition of a trace of sodic carbonate.
	30	305	
	50	300	
1	10	300	
4	10	300	Injected double dose.
4	30	310	
4	35	310	
4	40	310	
4	50	310	Injected double dose.
15		. . .	Killed by pithing.

Experiment No. 38.

Time: min. sec.		Pulsations per minute.	REMARKS.
Normal	. . .	270	Injected intravenously the *water-venom-globulin* from one
	10	290	minim of fresh venom of the *Crotalus adamanteus*.
	20	295	
	30	295	
	50	260	
1	00	255	
1	30	260	
3	30	265	
5	30	260	
7	30	265	
9	30	265	
12	30	260	
14	30	260	
16	30	270	
17	30	275	
19	30	275	
21	30	260	Clot in canula.
26	00	260	
28	00	260	
30	00	260	
30	15	260	Clot in canula.
35	00	260	
37	00	260	
39	00	255	Animal killed by pithing.

Experiment No. 39.

Time: min. sec.		Pulsations per minute.	REMARKS.
Normal	. .	280	Injected intravenously the *water-venom-globulin* from 0.004
	30	270	gram dried venom of the *Ancistrodon piscivorus* dissolved
	50	230	in 1 c. c. distilled water by the addition of a few crystals of
1	00	220 ?	sodic chloride.
1	30	280	
1	50	260	Injected a similar dose.
2	30	180	
2	40	260	
3	10	280	

Experiment No. 40.

Time: min. sec.	Pulsations per minute.	REMARKS.
Normal	312	Injected intravenously the *water-venom-globulin* from 0.015 gram dried venom of the *Crotalus adamanteus*.
0	...	
10	314	
20	...	
30	360	
1 30	316	
2 30	304	
5 30	324	
10 30	354	
14 30	360	Hæmaturia.
19 30	396	
24 30	384	
29 30	372	
34 30	372	
42 30	372	
47 30	372	
52 30	372	
57 30	360	
67 30	316	
77 30	316	
80 30	120	
85	...	Dead; ecchymoses in intestines; blood fluid.

Experiment No. 41.

Time: min. sec.	Pulsations per minute.	REMARKS.
Normal	280	Injected intravenously 0.0012 gram *copper-venom-globulin* from the dried venom of the *Crotalus adamanteus*.
10	285	
20	290	
30	285	
50	280	
2 50	270	
4 50	270	
6 50	260	
8 50	260	
10 50	255	
11 50	260	
17 20	280	
18 20	280	Injected a similar dose.
18 30	300	
18 40	290	
18 50	285	
19 00	280	
20 00	285	
22 00	285	
27 00	290	

Experiment No. 42.

Time: min. sec.		Pulsations per minute.	REMARKS.
Normal	...	290	Injected intravenously 0.00225 gram *copper-venom-globulin* from the dried venom of the *Crotalus adamanteus*.
	10	290	
	30	305	
1	00	310	
3	00	310	
5	00	310	
7	00	310	
8	00	310	Clot in canula.
10	00	310	
12	00	312	
22	00	280	
24	00	280	
26	00	280	Injected 0.0045 gram.
26	10	285	
26	20	290	
26	30	290	
27	00	290	
27	00	290	
29	00	280	
31	00	270	
34	00	250	
39	00	295	
41	00	290	
43	00	285	
45	00	285	
52	00	295	
58	00	295	

Experiment No. 43.

Time: min. sec.		Pulsations per minute.	REMARKS.
Normal	...	290	Injected intravenously 0.0017 gram *dialysis-venom-globulin* from the dried venom of the *Crotalus adamanteus* dissolved in 1 c. c. distilled water with a trace of sodic carbonate.
	20	365	
	40	305	
	50	305	
1	20	295	
3	20	275	
5	20	275	Animal broke loose.
18	20	280	
18	23	...	Injected 0.0034 gram *dialysis-venom-globulin*.
18	30	290	
18	45	280	
19	05	...	Struggles.
19	25	270	
19	55	270	
20	25	270	
21	25	138	
22	00	315	
30	00	...	Dead.

10 May, 1886.

Experiment No. 44.

Time: min. sec.		Pulsations per minute	REMARKS
Normal	. . .	265	Injected intravenously *dialysis-venom-globulin* from the dried venom of the *Crotalus adamanteus*.
	20	280	
	30	290	
1	00	270	
1	20	270	
1	40	270	
2	00	265	
2	40	280	
3	40	280	
5	40	285	
6	20	285	
6	50	285	
7	20	300	
7	50	300	
8	50	300	
9	20	300	
9	50	300	
10	50	300	
11	50	300	
12	50	300	
14	20	300	

Experiment No. 45.

Time: min. sec.		Pulsations per minute	REMARKS
Normal	. . .	270	Injected intravenously 0.0017 gram *dialysis-venom-globulin* from the dried venom of the *Crotalus adamanteus*.
	10	280	
	20	300	
	30	295	
1	00	280	
3	00	275	
5	00	260	
7	00	250	Clot formed in canula.
16	00	. . .	Injected 0.0034 gram.
17	30	255	" " "
17	40	275	
18	30	260	
20	30	270	Struggles.
23	30	260	
28	30	260	
43	30	270	
53	30	290	

The Actions of the Venom Globulins on Animals with Cut Pneumogastric Nerves.—In five experiments on animals with cut pneumogastric nerves—one with the *water-venom-globulin*, two with *copper-venom-globulin*, one with *dialysis-venom-globulin* (all from the *Crotalus adamanteus*), and one with the *water-venom-globulin* of the Cobra—there was a tendency to a lowered pulse-rate, although in one experiment there was a primary increase, and in another a slight increase above the healthy number after repeated injections. The effects were generally less than in normal animal.

THE ACTION OF VENOMS UPON THE PULSE-RATE.

Experiment No. 46.

Time: min. sec.	Pulsations per minute.	REMARKS.
Normal . . .	205	Pneumogastric nerves cut. Injected intravenously 0.0011
10	220	gram *water-venom-globulin* from the dried venom of the
20	230	*Crotalus adamanteus.*
1 00	210	
1 40	190	
3 40	170	
5 40	180	Clot in canula.

Experiment No. 47.

Time: min. sec.	Pulsations per minute.	REMARKS.
Normal . . .	324	Pneumogastric nerves cut. Injected intravenously *water-*
0	. . .	*venom-globulin* from 0.335 gram dried Cobra venom dis-
15	. . .	solved in 1 c. c. distilled water.
25	312	
45	318	
1 15	264	
2 00	300	
4 00	304	
8 00	276	
13 00	288	
18 00	310	
23 00	319	
28	. . .	Animal broke loose from canula.

Experiment No. 48.

Time: min. sec.	Pulsations per minute.	REMARKS.
Normal . . .	305	Pneumogastric nerves cut. Injected intravenously 0.0012
20	300	gram *copper-venom-globulin* from the dried venom of the
40	288	*Crotalus adamanteus.*
1 10	285	
3 10	285	
5 10	285	
7 10	300	
9 10	290	
23 10	310	Injected 0.0024 gram.
23 40	310	" " "
24 15	310	Killed.

Experiment No. 49.

Time: min. sec.	Pulsations per minute.	REMARKS.
Normal . . .	300	Pneumogastric nerves cut. Injected intravenously 0.0012
20	300	gram *copper-venom-globulin* from the dried venom of the
40	300	*Crotalus adamanteus.*
50	300	
2 50	300	
4 50	300	
6 50	300	
8 50	300	
11 50	300	

Time: min. sec.	Pulsations per minute.	REMARKS.
13 50	300	
15 50	300	Injected 0.0024 **gram**.
16 10	300	
16 20	300	" " "
16 30	300	
16 45	300	
17 45	300	**Struggles.**
19 45	300	
21 **45**	300	
23 45	270	
25 45	270	
26 45	270	
27 **00**	270	Animal killed by pithing.

Experiment No. 50.

Time: min. sec.	Pulsations per minute.	REMARKS.
Normal . .	310	Pneumogastric nerves cut. Injected intravenously 0.0017
10	305	gram *dialysis venom-globulin* from the dried venom of the
20	300	*Crotalus adamanteus*.
30	300	
50	310	Struggles.
1 50	300	
4 20	310	
6 20	300	
8 20	295	
10 20	295	
12 **20**	**300**	**Struggles.**
17 **50**	**300**	
18 20	300	Injected 0.0034 gram.
18 40	310	" " "
19 00	. . .	Struggles.
19 15	320	" .
19 20	330	"
21 50	320	
23 00	310	
25 00	310	
27 00	310	
29 00	310	
34 00	305	
34 30	300	
38 30	290	
41 00	280	
47 00	217	
49 00	. . .	Dead.

The Actions of Venom Globulins on the Pulse-rate in Animals with the Pneumogastric Nerves and Cervical Spinal Cord Cut.—In four experiments in which the pneumogastric nerves and spinal cord in the middle cervical region were cut—one was made with the *water-venom-globulin*, one with the *copper-venom-globulin*, and two with *dialysis-venom-globulin* of the *Crotalus adamanteus*: in one experiment

THE ACTION OF VENOMS UPON THE PULSE-RATE. 77

there was a fall followed by a rise to the normal, and succeeded by a slight fall; in a second the pulse-rate always remained below normal, while in the third there was an almost inappreciable rise, this followed by a fall, and by an increase due to a further injection of the poison. The last showed a slight fall, then a return to the normal.

Experiment No. 51.

Time: min. sec.	Pulsations per minute.	REMARKS.
Normal . . .	250	Pneumogastric nerves and cord cut. Injected intravenously
10	250	0.0011 gram *water-venom-globulin* from the dried venom
30	215	of the *Crotalus adamanteus*.
1 00	240	
1 10	240	
2 10	245	
3 10	245	
5 10	245	
7 10	245	
9 10	245	
11 10	250	
15 10	250	
17 40	**245**	
19 00	240	
21 00	240	
23 00	240	
25 00	240	
27 00	240	Killed.

Experiment No. 52.

Time: min. sec.	Pulsations per minute.	REMARKS.
Normal . . .	255	Pneumogastric nerves and cord cut. Injected intravenously
10	**255**	0.0048 gram *copper-venom-globulin* from the dried venom
20	**240**	of the *Crotalus adamanteus*.
40	250	
1 00	250	
3 30	240	
5 30	225	
9 30	225	
11 30	210	
13 30	210	
16 30	210	
17 00	204	Injected 0.0048 gram.
17 30	210	
18 00	210	
20 00	210	
24 00	195	
26 00	180	
28 00	180	
30 00	180	
32 00	180	
34 00	187	Killed.

Experiment No. 53.

Time: min. sec.		Pulsations per minute.	REMARKS.
Normal	. . .	240	Pneumogastric nerves and cord cut. Injected intravenously
	10	240	0.0017 gram *dialysis-venom-globulin* from the dried venom
	30	245	of the *Crotalus adamanteus*.
1	00	230	
3	00	220	
6	00	220	
8	30	220	
10	30	230	Injected 0.0034 gram.
12	30	230	" " "
12	50	220	
13	10	225	
13	30	210	
13	50	210	
15	00	200	
16	00	190	
18	00	. . .	Dead.

Experiment No. 54.

Time: min. sec.		Pulsations per minute.	REMARKS
Normal	. . .	300	Pneumogastric nerves and cord cut. Injected intravenously
	10	290	0.0068 gram *dialysis-venom-globulin* from the dried venom
	20	290	of the *Crotalus adamanteus*.
	30	300	
	40	300	Tremors.
	50	300	
1	00	300	
1	10	300	
1	20	300	Clot formed in canula.
9	00	. . .	Dead.

A review of the results of these experiments with the globulins on the pulse-rate in normal animals indicates that water-venom-globulin is the most potent, and the copper-venom-globulin the least so. With the former there occurred in four of the six experiments a primary increase followed by a fall, while in the other two there was a diminution from the first. In experiments with the copper-venom-globulin and dialysis-venom-globulin there was always a primary increase, and in four out of the five experiments this was followed by a decline.

After section of the pneumogastric nerves a primary increase (due probably to some accidental cause) occurred in one out of the five experiments, in two of the other four there at first was no appreciable change, and then a diminution, while in the remaining two there was a lessening of the rate from the time of injection. These results suggest that the increase of the pulse-rate, which occurred in animals with intact vagi, was in some degree at least dependent upon an influence exerted through the pneumogastric centres and nerves. It will be observed that we here have results which are directly opposed to what we have seen with pure venom; that is a lessened tendency to the primary increase of the

THE ACTION OF VENOMS UPON THE PULSE-RATE.   79

pulse after section of the pneumogastric nerves. If the increase in the pulse-rate in normal animals is due for the most part to excitation of the accelerator centres, whereby impulses are generated which pass chiefly through the accelerator fibres running in the spinal cord, it would seem probable that the accelerator impulses induced by the globulins take for the most part the course of the fibres through the pneumogastric nerves, but are much feebler than the impulses which are generated by the pure venoms, and which take their path chiefly through the fibres in the spinal cord.

After section of both the pneumogastric nerves and cervical spinal cord, we found in all of our experiments a diminution in the heart-beats; this must be due to a direct action of the globulin upon the heart.

It therefore seems probable that the globulins cause a primary increase of the pulse by an excitation of the accelerator centres, whereby impulses are conveyed principally by the accelerator fibres passing through the pneumogastric nerves; and a diminution of the heart beats by a direct action on the heart.

SECTION III.—THE ACTIONS OF VENOM PEPTONES UPON THE PULSE-RATE

The Action of Venom Peptones on the Pulse-rate.—In seven experiments made with peptone on normal animals—four with the peptone from the venom of the *Crotalus adamanteus*; one with that of the *Ancistrodon piscivorus*; and two with that of the *Cobra*—we find results which vary and which resemble closely those obtained by the administration of pure venom. In three experiments there was a primary increase of pulse followed generally by a diminution; in three the pulse remained below normal; while with *Ancistrodon* peptone there was a primary fall of rate followed by a rise.

The differences in the results, as in previous experiments, do not seem to depend at all upon the dose or the variety of venom from which the peptone was obtained.

Experiment No. 55.

Time : min. sec.		Pulsations per minute.	REMARKS
Normal		280	Injected intravenously the *peptone* from 0.015 gram dried venom of the *Crotalus adamanteus*.
	20	102	
	30	190	
	40	190	
	50	190	
1	00	180	
3	00	190	
6	00	340	Struggles.
11	00	285	Struggles. Broke loose.
49	00	...	Dead.

Experiment No. 56.

Time: min. sec.	Pulsations per minute.	REMARKS
Normal . . .	225	Injected intravenously the *peptone* from 0.03 gram **dried**
10	260	venom of the *Crotalus adamanteus*.
30	260	
1 00	285	
2 00	270	
5 00	260	
9 00	250	**Killed by pithing.**

Experiment No. 57.

Time: min. sec.	Pulsations per minute.	**REMARKS.**
Normal . . .	280	**Injected intravenously the *peptone* from 0.015** gram dried
30	280	**venom of the *Crotalus adamanteus*.**
1 00	270	
1 30	285	
4 30	270	
10 30	270	Injected double the amount.
10 50	270	
11 10	270	
11 30	270	

Experiment No. 58.

Time: min. sec.	Pulsations per minute.	REMARKS.
Normal . . .	270	Injected intravenously **the** *peptone* from **0.015** gram dried
10	270	**venom of** the *Crotalus adamanteus*.
20	**270**	
30	**270**	
1 00	280	
1 30	290	
2 00	290	
2 30	290	
3 00	260	
5 00	260	Injected a similar quantity.
5 20	260	
5 40	260	
6 00	260	
6 30	255	
7 00	250	Injected double the quantity.
7 30	240	
8 00	260	

Experiment No. 59.

Time: min. sec.	Pulsations per minute.	REMARKS.
Normal . . .	300	Injected intravenously the *peptone* from 0.05 gram dried
10	175	venom of the *Ancistrodon piscivorus*.
30	110	
1 00	285	
1 20	290	
1 40	300	
2 00	310	

THE ACTION OF VENOMS UPON THE PULSE-RATE. 81

Time: min. sec.	Pulsations per minute	REMARKS
2 20	340	
2 40	315	Injected *peptone* from 0.01 gram venom.
5 40	340	
5 50	348	
6 00	340	
6 20	340	
6 30	340	
13 30	...	Killed.

Experiment No. 60.

Time: min. sec.	Pulsations per minute	REMARKS
Normal ...	225	Injected intravenously the *peptone* from 0.005 gram dried Cobra venom.
10	220	
20	220	
30	220	
40	230	
1 00	230	
1 20	220	
3 20	204	Struggles.
6 20	165	
11 20	280	Convulsive twitchings.
16 20	260	
16 50	200	
17 50	150	
18 20	100	Blood is asphyxiated; no respiration.
18 50	...	
19 00	130	
19 10	190	
19 30	110	
19 50	38	
20 10	35	
20 30	35	Killed.

Experiment No. 61.

Time: min. sec.	Pulsations per minute	REMARKS
Normal ...	290	Injected intravenously the *peptone* from 0.06 gram dried Cobra venom.
10	300	
15	310	
30	240	
50	245	
3 20	225	
8 20	97	Tonic convulsions.
10 20	72	Convulsive twitchings; asphyxiated blood; respiration ceased.
10 40	300	
10 50	300	
11 20	105	
11 40	85	
12 00	120	
12 20	90	
12 40	...	Dead.

11 May, 1880.

The Actions of Venom Peptones on Animals in which the Pneumogastric Nerves had been Cut.—Four experiments were made with the peptone on animals the pneumogastric nerves of which had been previously cut. In three of the four there was a primary increase in the pulse, while in the fourth there was a temporary diminution followed by a rise above normal. These experiments are in accord with those made with pure venom, and indicate a greater tendency to primary pulse frequency **after** section of the pneumogastric nerves.

One of the above experiments was made with the peptone from the *Crotalus adamanteus*; one with the *Ancistrodon piscivorus*; and two with the Cobra.

Experiment No. 62.

Time: min. sec.	Pulsations per minute.	REMARKS.
Normal . . .	285	Pneumogastric nerves cut. Injected intravenously the *peptone* from 0.015 gram dried venom of the *Crotalus adamanteus*.
10	285	
30	285	
40	300	Struggles.
50	300	
1 00	300	
1 20	300	
1 30	300	
3 30	315	
8 30	330	Struggles.
15 30	345	
20 30	325	
25 30	315	
30 30	315	
35 00	315	
40 00	315	
45 00	270	
50 00	255	
62 00	285	Killed.

Experiment No. 63.

Time: min. sec.	Pulsations per minute.	REMARKS.
Normal . . .	295	Pneumogastric nerves cut. Injected intravenously the *peptone* from 0.015 gram dried venom of the *Ancistrodon piscivorus*.
30	315	
40	315	
50	310	
1 00	310	
1 20	315	
1 40	320	
2 00	330	
2 30	310	
2 40	102	
2 50	150	
3 00	240	
4 20	. . .	Dead.

THE ACTION OF VENOMS UPON THE PULSE-RATE.

Experiment No. 64.

Time: min. sec.		Pulsations per minute.	REMARKS.
Normal	. . .	230	Pneumogastric nerves cut. Injected intravenously the *peptone* from 0.005 gram dried Cobra venom.
	10	235	
	20	240	
	40	230	
1	00	230	
1	30	230	
3	30	230	
5	30	230	
7	30	230	
9	30	220	
10	00	220	
21	30	225	Twitchings.
25	30	230	
26	30	235	Killed.

Experiment No. 65.

Time: min. sec.		Pulsations per minute.	REMARKS.
Normal	. . .	265	Pneumogastric nerves cut. Injected intravenously the *peptone* from 0.006 gram dried Cobra venom.
	10	265	
	20	255	
	40	260	
1	00	260	
3	00	270	
5	00	270	
16	00	275	
34	00	. . .	Dead.

The Actions of Venom Peptones upon the Pulse-rate of Animals after Section of the Pneumogastric Nerves and Cervical Spinal Cord.—Six experiments made on animals in which the heart was cut off from central nervous influence by section of the pneumogastric nerves and section of the spinal cord in the middle cervical region gave uniform results. Three were made with peptone from the venom of the *Crotalus adamanteus*; two with the peptone from the *Ancistrodon piscivorus*, and one with that of the *Cobra*. In all of these experiments there was a diminution of the pulse-rate, and usually this was well marked. These results are also in accord with what was found in experiments with pure venom.

Experiment No. 66.

Time: min. sec.		Pulsations per minute.	REMARKS.
Normal	. . .	200	Pneumogastric nerves and cord cut. Injected intravenously the *peptone* from 0.015 gram dried venom of the *Crotalus adamanteus*.
	10	185	
	20	195	
	40	195	
1	00	190	
3	00	160	
13	00	195	
16	00	200	

Time: min. sec.	Pulsations per minute.	REMARKS
17 00	190	Injected peptone from 0.03 gram dried venom.
17 30	...	
19 30	187	
21 30	190	
23 30	180	
25 30	170	
31 30	165	
34 30	...	Dead.

Experiment No. 67.

	Time: min. sec.	Pulsations per minute.	REMARKS.
Normal	...	282	Pneumogastric nerves and cord cut. Injected intravenously the peptone from 0.015 gram dried venom of the *Crotalus adamanteus*.
	0	...	
	15	276	
	20	...	
	30	264	
1	00	264	
1	30	246	
2	00	234	

Experiment No. 68.

	Time: min. sec.	Pulsations per minute.	REMARKS.
Normal	...	324	Pneumogastric nerves and cord cut. Injected intravenously the peptone from 0.015 gram dried Cobra venom.
	0	...	
	11	300	
	40	318	
1	40	276	
5	40	272	
10	40	276	
15	40	306	
16	00	...	} Injected a similar dose.
16	06	...	
16	15	306	
16	25	300	
17	00	276	
18	00	270	
23	00	...	Dead.

In the above series of experiments with venom peptones we find results which agree with those in which the pure venoms were used. We conclude, therefore, that the peptones cause a primary increase and a secondary diminution of the pulse-rate, and that they occasion primary hastening of the heart beat by excitation of the accelerator centres in the medulla, and that the impulses are carried through fibres passing chiefly by the spinal cord. This increase is more marked after section of the pneumogastric nerves, thus suggesting that this principle has some direct or indirect effect upon the pneumogastric centres, tending to slow the action of the heart and to neutralize the accelerator influence. Peptones cause the diminution of the heart beat by a direct action on that organ.

CHAPTER VIII.

THE ACTION OF VENOMS AND THEIR ISOLATED GLOBULINS AND PEPTONES UPON THE ARTERIAL PRESSURE.

SECTION I.—PURE VENOM.

THE experiments made on the blood pressure with venoms and their isolated poisons were all made on rabbits. The manometer tube was connected with one of the carotid arteries, and the injections were always made into the external jugular vein unless otherwise noted.

Eighteen experiments were performed with the venoms of different species of serpents, and in all of them there was a distinct lowering of the blood-pressure. It fell immediately after the injection, and indeed sometimes before injection was complete, and the fall was generally so marked as to indicate a most profound action of the poison upon some part or parts of the circulatory apparatus. If the dose be not immediately fatal the pressure gradually rises, but finally undergoes a more or less steady decline to death. At other times the pressure sinks without subsequent rise until death ensues.

The tendency in Cobra poisoning is to a decided rise of pressure following the primary fall. In five out of six experiments with this venom the primary fall was followed by a rise which went above the normal.

Of the eighteen experiments, five were made with the venom of *Crotalus adamanteus*, in two of which the poison was given hypodermatically; two with that of the *Crotalus horridus*; two with the venom of *Ancistrodon piscivorus*; one each with the poisons *Ancistrodon contortrix*, *Crotalophorus miliarius*, and *Daboia Russellii*; and six with the venom of the Cobra. In all cases ether was given freely to the animal poisoned.

Action of the Pure Venoms upon the Arterial Pressure in Normal Animals.

Experiment No. 1.

	Time : min. sec.	Pressure m. m.	REMARKS.
Normal	. . .	126	Injected into the thigh of a large rabbit 1 drop of fresh venom from the *Crotalus adamanteus*.
	20	126	
	40	126	
	1 00	124	
	1 20	122	
	1 40	120	
	2 00	118	Clot formed in canula.
	5 00	114	

Time: min. sec.	Pressure m. m.	REMARKS
7 00	84	
8 00	84	
9 00	84	
10 00	82	
11 00	82	
12 00	78	
13 00	82	
20 00	52	
21 30	56	
23 00	64	
25 00	56	
26		Struggles followed by death.

Experiment No. 2.

Time: min. sec.	Pressure m. m.	REMARKS
Normal . . .	144	Injected into the thigh of a rabbit 3 drops of fresh venom from the *Crotalus adamanteus*.
20	144	
40	138	
1 00	136	
1 20	130	
1 40	130	
2 00	126	Struggles. Animal tore loose from the canula.

Experiment No. 3.

Time: min. sec.	Pressure m. m.	REMARKS
Normal . . .	70	**Injected intravenously 0.003** gram dried venom of the *Crotalus adamanteus* in 1 c. c. distilled water.
10	58	
20	54	
30	60	
40	68	
50	68	
1 00	68	
1 20	68	
1 40	62	
2 00	58	
2 20	54	
2 40	44	Death.

Experiment No. 4.

Time: min. sec.	Pressure m. m.	REMARKS
Normal . . .	140	Injected intravenously 0.003 gram dried venom of the *Crotalus adamanteus* dissolved in 1 c. c. distilled water.
5	76	
10	72	
20	64	
30	62	
40	60	
1 00	58	
1 20	58	
1 40	56	
2 10	54	

Time: min. sec.	Pressure m. m.	REMARKS
4 00	45	
7 00	40	
8 00	110	Struggles.
8 10	94	
8 20	74	
8 30	78	
8 40	76	
8 50	60	
10 30	56	
13 00	64	
13 30	...	Injection as before.
13 50	50	
14 00	50	
16 00	...	" "
16 05	48	
16 30	44	
17 00	...	Dead. Heart in complete diastole. Ecchymoses in pericardium and peritoneum. Blood incoagulable.

Experiment No. 5.

Time: min. sec.	Pressure m. m.	REMARKS
Normal ...	124	Injected intravenously 0.015 gram dried venom of the Crotalus adamanteus dissolved in 1 c. c. distilled water.
10	100	
20	60	Struggles.
30	96	
40	84	
1 00	70	
1 20	56	
1 40	48	
1 55	44	
4 00	38	Pulse feeble.
7 00	32	
8 00	...	Dead.

Experiment No. 6.

Time: min. sec.	Pressure m. m.	REMARKS
Normal ...	104	Injected intravenously 0.015 gram dried venom of the Crotalus horridus dissolved in 1 c. c. distilled water
5	84	
10	68	
20	74	
30	80	
40	78	
50	70	
1 00	64	
1 10	60	
1 20	56	
1 30	52	
1 40	48	
3 40	40	
5 40	44	
7 40	46	
9 40	42	
10 10	38	Convulsions. Dead. Some ecchymoses; blood fluid.

Experiment No. 7.

Time: min. sec.	Pressure m. m.	REMARKS.
Normal ...	110	Injected intravenously 0.015 gram dried venom of the *Crotalus horridus* dissolved in 1 c. c. distilled water.
5	76	
10	76	
20	78	
30	70	Animal broke loose from mouth-piece, and was firmly held and refixed.
2 40	60	
3 00	44	
5 00	42	
6 30	40	
7 30	36	
8 00	32	
8 30	26	
9 00	20	
10 00	10	**Dead.** Respiration failed before the heart.

Experiment No. 8.

Time: min. sec.	Pressure m. m.	REMARKS.
Normal ...	138	Injected intravenously 0.004 gram dried venom of the *Ancistrodon piscivorus* dissolved in 1 c. c. distilled water.
20	80	
30	64	Convulsions.
40	74	
1 00	84	
1 30	76	Injected as above.
1 50	60	Struggles.
2 10	**74**	
2 30	74	
3 00	84	
4 00	...	Killed by pithing.

Experiment No. 9.

Time: min. sec.	Pressure m. m.	REMARKS.
Normal ..	134	Injected intravenously 0.004 gram dried venom of the *Ancistrodon piscivorus* dissolved in 0.5 c. c. distilled water.
10	108	
20	72	
30	70	
1 00	70	
1 30	72	
2 00	74	
2 30	70	
3 00	70	
5 00	70	
5 30	68	Injection repeated as before
5 35	76	
5 45	70	
6 05	60	
6 15	62	Convulsive movements.
6 25	66	
6 45	60	
7 05	52	

THE ACTION OF VENOMS UPON ARTERIAL PRESSURE.

Time: min. sec.	Pressure m. m.	REMARKS.
7 15	54	
7 25	72	
7 35	70	
7 45	66	
7 55	104	
8 05	120	
8 15	100	
8 25	86	Animal died in a few minutes.

Experiment No. 10.

	Time: min. sec.	Pressure m. m.	REMARKS.
Normal	. . .	96	Injected intravenously 0.003 gram dried venom of the *Ancistrodon contortrix* dissolved in 1 c. c. distilled water.
	10	70	
	30	76	
	1 00	72	
	1 30	72	
	2 00	72	
	2 30	70	
	4 30	48	
	5 30	48	Injection repeated.
	5 50	42	Struggles.
	6 30	50	
	6 50	50	
	7 50	46	Injection repeated.
	10 00	. . .	Dead. Heart flabby; blood incoagulable; no ecchymoses.

Experiment No. 11.

	Time: min. sec.	Pressure m. m.	REMARKS.
Normal	. . .	170	Injected intravenously 0.003 gram dried venom of the *Crotalophorus miliarius* dissolved in 1 c. c. distilled water.
	20	122	
	30	120	
	40	136	
	50	116	
	1 00	150	
	1 10	84	
	2 10	108	
	2 40	106	Injection repeated as before.
	2 50	80	
	3 20	84	
	5 20	70	
	7 20	74	
	9 20	80	Killed by pithing.

Experiment No. 12.

	Time: min. sec.	Pressure m. m.	REMARKS.
Normal	. . .	130	Injected intravenously 0.003 gram dried Cobra venom dissolved in 1 c. c. distilled water.
	10	110	
	20	102	
	30	96	
	40	88	

12 June, 1886.

Time: min. sec.	Pressure m. m.	REMARKS
50	70	
1 00	78	
1 10	70	
1 20	52	
1 30	52	Convulsions.
1 40	50	
2 00	38	
2 20	34	
3 20	...	Dead. Heart in systole; blood clotted; no ecchymoses.

Experiment No. 13.

	Time: min. sec.	Pressure m. m.	REMARKS
Normal	...	146	Injected intravenously 0.003 gram dried Cobra venom dissolved in 1 c. c. distilled water.
	10	150	
	30	140	
	1 00	134	
	1 20	134	
	3 20	140	
	5 20	148	Death from hemorrhage; artery torn.

Experiment No. 14.

	Time: min. sec.	Pressure m. m.	REMARKS
Normal	...	132	Injected intravenously 0.003 gram dried Cobra venom dissolved in 1 c. c. distilled water with a few crystals of sodic chloride.
	1 00	122	
	3 00	120	
	8 00	130	
	10 00	138	
	15 00	106	
	17 00	66	
	18 00	48	
	19 00	...	Dead

Experiment No. 15.

	Time: min. sec.	Pressure m. m.	REMARKS
Normal	...	145	Injected intravenously 0.003 gram dried venom of the *Cobra* dissolved in 1 c. c. distilled water.
	20	142	
	40	135	
	1 00	128	
	1 30	130	
	2 00	140	
	4 00	150	
	9 00	143	
	12 00	138	
	14 00	158	Respiration ceased; artificial respiration used.
	20 00	135	

THE ACTION OF VENOMS UPON ARTERIAL PRESSURE.

Experiment No. 16.

	Time: min. sec.	Pressure m. m.	REMARKS
Normal	. . .	120	Injected intravenously 0.005 gram dried Cobra venom dissolved in 1 c. c. distilled water with a little sodic chloride and filtered.
	20	108	
	30	96	
	40	88	
	1 00	90	
	1 20	82	
	1 40	96	
	4 40	94	
	7 40	100	A clot was probably beginning to form in the canula, and no dependence is to be placed upon the after record.
	8 40	122	
	9 10	156	Struggles. Clot in canula.
	15 00	. . .	Dead.

Experiment No. 17.

	Time: min. sec.	Pressure m. m.	REMARKS
Normal	. . .	90	Injected intravenously 0.015 gram dried Cobra venom dissolved in 1 c. c. distilled water.
	10	94	
	20	66	
	30	84	
	40	90	
	1 00	88	
	1 30	86	
	2 00	78	
	2 10	74	
	4 10	90	
	4 40	96	
	5 10	92	
	6 10	72	
	6 20	60	
	6 50	32	
	7 30	24	
	8 20	6	Dead.

Experiment No. 18.

	Time: min. sec.	Pressure m. m.	REMARKS
Normal	. . .	110	Injected intravenously 0.005 gram dried venom of the *Daboia Russellii* dissolved in 0.5 c. c. distilled water.
	10	112	
	11	. . .	Pressure falling.
	15	80	
	20	64	
	30	46	
	40	30	Violent general convulsions.
	50	54	
	1 00	46	
	1 15	46	
	2 00	. . .	Dead. Heart in diastole; no ecchymoses; after twenty-four hours the blood is still fluid. Artificial respiration was used in this experiment from the beginning.

This single experiment confirms the statements of Fayrer and of Wall in regard to the convulsivant power of Daboia. The spasms are not due to defect of oxygen, as they arise early and occur despite the use of artificial respiration. Ancistrodon venom seems to have the same capacity to produce convulsions.

The Action of Pure Venoms on the Blood Pressure of Animals with Cut Pneumogastric Nerves.—After section of the pneumogastric nerves, including the depressor fibres, we find that the same alterations occur in the blood pressure as in normal animals. Nine experiments were made altogether: two with the venom of the *Crotalus adamanteus*; one with the *Crotalus horridus*; two with the *Ancistrodon piscivorus*; one with the *Ancistrodon contortrix*; and three with the Cobra.

Experiment No. 19.

	Time: min. sec.	Pressure m. m.	REMARKS.
Normal	. . .	96	Injected intravenously 0.003 gram dried venom of the *Crotalus adamanteus* dissolved in 1 c. c. distilled water.
	10	74	
	30	68	
	1 00	68	
	1 30	76	
	2 00	64	
	2 30	60	
	5 30	44	
	7 00	44	
	8 30	48	Injected the same as above.
	8 40	42	
	9 00	46	
	9 10	44	
	9 20	44	
	9 30	44	Dead.

Experiment No. 20.

	Time: min. sec.	Pressure m. m.	REMARKS.
Normal	. . .	130	Injected intravenously 0.015 gram dried venom of the *Crotalus adamanteus* dissolved in 1 c. c. distilled water.
	10	100	
	20	90	
	30	96	
	40	76	
	50	56	
	1 00	50	
	1 10	38	
	1 20	32	
	1 30	22	Dead.

THE ACTION OF VENOMS UPON ARTERIAL PRESSURE.

Experiment No. 21.

Time: min. sec.	Pressure m. m.	REMARKS
Normal . . .	144	Injected intravenously 0.015 gram dried venom of the *Crotalus horridus* dissolved in 1 c. c. distilled water.
10	146	Struggles.
20	124	Violent struggles.
40	124	
1 00	94	
1 20	80	
1 40	70	
2 00	56	
2 20	54	
4 20	54	
5 50	44	
8 00	12	Dead.

Experiment No. 22.

Time: min. sec.	Pressure m. m.	REMARKS
Normal . . .	90	Injected intravenously 0.0635 gram dried venom of the *Ancistrodon piscivorus* dissolved in 1 c. c. distilled water.
8	76	
10	54	
20	44	
30	50	Convulsions.
40	54	
1 00	44	
1 20	24	
1 40	. . .	Dead.

Experiment No. 23.

Time: min. sec.	Pressure m. m.	REMARKS
Normal . . .	110	Injected intravenously 0.003 gram dried venom of the *Ancistrodon piscivorus* dissolved in 1 c. c. distilled water.
10	84	
20	64	
30	68	
40	78	Struggles.
50	76	
1 00	72	
1 20	66	
3 50	52	
5 50	64	Struggles.
6 00	86	Convulsions.
8 00	100	
10 00	74	
12 00	70	
14 00	66	
19 00	70	Injection repeated, using the same amount.
19 10	70	
19 20	52	
19 30	98	
19 40	68	
19 50	90	

Time: min. sec.	Pressure m. m.	REMARKS.
20 00	90	
20 20	72	
21 20	60	
23 20	56	
25 20	60	
28 20	70	
33 20	70	
38 20	66	
44 20	58	Third injection, same amount.
44 30	50	
44 40	48	
44 50	44	
45 00	36	
45 10	26	Dead.

Experiment No. 24.

	Time: min. sec.	Pressure m. m.	REMARKS
Normal	. . .	154	Injected intravenously 0.003 gram dried venom of the *Ancistrodon contortrix* dissolved in 1 c. c. distilled water.
	10	114	
	20	86	
	30	76	
	40	82	
	1 00	84	
	1 50	80	
	4 20	98	
	7 00	100	**Injection repeated.**
	7 05	138	**Struggles.**
	7 10	110	
	7 20	106	
	7 40	108	
	8 00	98	
	8 30	88	
	9 00	80	
	11 30	78	Third injection.
	11 40	76	
	12 00	70	
	12 30	66	
	13 00	68	
	13 30	68	
	19 00	72	Fourth injection.
	19 20	54	
	19 40	50	
	20 20	44	
	21 50	40	
	22 50	38	Dead. Heart in diastole; blood remains fluid; muscles all respond to electrical irritation; motor nerves react feebly.

Experiment No. 25.

Time: min. sec.	Pressure m. m.	REMARKS
Normal . . .	148	Injected intravenously 0.003 gram dried Cobra venom dissolved in 1 c. c. distilled water with a few crystals of sodic chloride and filtered.
10	146	
30	136	
1 00	140	
1 30	140	
3 30	150	
6 30	150	
10 30	146	
14 30	152	
16 30	156	Clot formed in canula. Animal killed.

Experiment No. 26.

Time: min. sec.	Pressure m. m.	REMARKS
Normal . . .	134	Injected intravenously 0.006 gram dried Cobra venom prepared as in the foregoing experiment.
10	136	
30	126	
1 00	118	
1 30	118	
3 30	132	
5 30	136	
7 30	136	
9 30	136	
11 30	136	
18 30	188	Convulsive movements; asphyxia; respiration ceased in three minutes.

Experiment No. 27.

Time: min. sec.	Pressure m. m.	REMARKS
Normal . . .	130	Injected intravenously 0.003 gram dried Cobra venom dissolved in 1 c. c. distilled water.
0	. . .	
10	. . .	
20	130	
1 00	118	
2 00	118	
4 00	115	
10	. . .	Clot in canula.
15	. . .	Dead from asphyxia

The Action of Pure Venoms on the Blood Pressure of Animals in which the Cervical Spinal Cord had been Divided.—Upon section of the spinal cord in the upper cervical region, by which the influence of the vaso-motor centres in the medulla is practically destroyed, the primary fall of pressure from venom is generally very slight, and after this diminution there is a secondary rise which may go above the normal. In one experiment with *Crotalus adamanteus* venom there was a rise of pressure for a moment at the time of injection; in one experiment with *Crotalus horridus*, in which a somewhat larger dose was used than in the others, there was a distinct rise of pressure a few seconds after injection, followed by a fall; and in the experiment with the Cobra the pressure never went below the normal, but in a few moments a rise occurred which continued to increase for half an hour, when the animal was killed.

In this series we observed a marked difference from the preceding (unless the dose had been immediately toxic), since the profound primary fall of pressure was not observed, excepting in a very slight degree if at all; we found, however, that the ultimate fall of pressure still occurred, save in the case of the Cobra.

Eight experiments were made: two with *Crotalus adamanteus*; one with *Crotalus horridus*; three with *Ancistrodon piscivorus*; one with *Ancistrodon contortrix*, and one with *Cobra* venom.

Experiment No. 28.

Time: min. sec.	Pressure m. m.	REMARKS.
Normal . . .	62	Injected intravenously 0.003 gram dried venom of the *Crotalus adamanteus* dissolved in 1 c. c. distilled water.
9	70	
20	56	
30	56	
40	58	
1 00	58	
1 20	56	
1 30	48	
1 40	46	
2 00	44	
2 20	40	
2 40	38	
5 20	36	
6 50	36	
7 20	36	
8 00	. . .	Dead.

Experiment No. 29.

Time: min. sec.	Pressure m. m.	REMARKS.
Normal . .	66	Injected intravenously 0.006 gram dried venom of the *Crotalus adamanteus* dissolved in 1 c. c. distilled water.
5	60	
10	58	
20	58	
30	64	
40	62	
50	58	
1 00	56	
1 30	48	
2 30	40	
3 30	60	
4 30	56	
5 30	40	
8 00	. . .	Dead.

THE ACTION OF VENOMS UPON ARTERIAL PRESSURE. 97

Experiment No. 30.

Time: min. sec.	Pressure m. m.	REMARKS.
Normal . . .	30	Injected intravenously 0.015 gram dried venom of the *Crotalus horridus* dissolved in 1 c. c. distilled water.
20	30	
40	46	
, 00	42	
1 20	38	
3 20	26	
5 20	26	
7 20	26	
9 20	24	
10 50	30	
12 50	30	
14 50	28	
16 50	26	
17 05	26	
17 35	10	Conjunctival reflexes gone.
18 00	. . .	Dead.

Experiment No. 31.

Time: min. sec.	Pressure m. m.	REMARKS.
Normal . . .	56	Injected intravenously 0.007 gram dried venom of the *Ancistrodon piscivorus* dissolved in 1 c. c. distilled water.
10	50	
20	46	
30	46	
40	40	
50	34	
1 00	30	
3 00	22	
5 00	28	
7 00	26	
10 00	18	Dead. The cord proved not to have been completely cut—a few fibres of the posterior columns remaining undivided.

Experiment No. 32.

Time: min. sec.	Pressure m. m.	REMARKS.
Normal . . .	58	Injected intravenously 0.003 gram dried venom of the *Ancistrodon piscivorus* dissolved in 1 c. c. distilled water.
10	54	
20	40	
30	32	Convulsions.
40	26	
50	16	Dead.

Experiment No. 33.

Time: min. sec.	Pressure m. m.	REMARKS.
Normal . . .	46	Injected intravenously 0.003 gram dried venom of the *Ancistrodon piscivorus* dissolved in 1 c. c. distilled water.
10	38	
20	40	
30	38	
40	40	
56	48	
3 00	38	
5 00	30	
8 00	28	Dead.

Experiment No. 34.

Time: min. sec.	Pressure m. m.	REMARKS.
Normal . . .	52	Injected intravenously 0.003 gram dried venom of the *Ancistrodon contortrix* dissolved in 1 c. c. distilled water.
10	44	
20	46	
40	50	
1 00	54	
3 00	46	
5 00	36	
7 00	34	
9 00	34	
11 00	30	
13 00	30	
15 00	30	
17 00	30	
19 00	30	
21 00	30	
23 00	30	
25 00	32	
27 00	32	
29 00	34	
56 00	34	
61 00	32	
75 00	30	Killed by pithing.

Experiment No. 35.

Time: min. sec.	Pressure m. m.	REMARKS.
Normal . . .	42	Injected intravenously 0.003 gram dried venom of the Cobra dissolved in 1 c. c. distilled water with a few crystals of sodic chloride and filtered.
10	46	
30	44	
1 00	42	
3 00	46	
6 00	48	
9 00	46	
12 00	50	
15 30	50	
18 30	52	
21 30	54	
24 30	56	
27 30	54	Injected the same as the foregoing.
27 40	56	
28 00	64	
28 30	68	
30 30	78	Clot formed in canula. Killed animal by pithing.

The Action of Pure Venoms on the Blood Pressure of Animals in which the Pneumogastric, Depressor, and Sympathetic Nerves and Spinal Cord have been Severed.—Since we found in the last series of experiments that after section of the cord there did not occur such a decided primary fall of pressure, it seemed obvious that the fall of pressure must be due, in major part at least, to a toxic depression

THE ACTION OF VENOMS UPON ARTERIAL PRESSURE 99

of the vaso-motor centres. A fall of pressure does, however, ultimately occur, and, excepting in the case of the Cobra, increases until death ensues.

In seven other experiments, supplementary to the above, in which we made section of the pneumogastric, depressor, and sympathetic nerves in the neck, and section of the spinal cord in the middle or upper cervical region, thus cutting off both the heart and capillaries from centric nervous influence, we obtained results which were practically the same.

Three of these experiments were made with the venom of the *Crotalus adamanteus*, one with that of the *Crotalus horridus*; one with the *Ancistrodon piscivorus*; one with the *Ancistrodon contortrix*, and one with the Cobra.

Experiment No. 36

	Time min. sec.	Pressure m. m.	REMARKS.
Normal	. .	62	Injected intravenously 0.003 gram dried venom of the *Crotalus adamanteus* dissolved in 1 c. c. distilled water.
	10	56	
	20	46	
	30	56	
	40	52	
	1 00	46	
	1 20	40	
	1 40	36	
	2 00	30	
	2 20	24	Dead. Heart arrested in diastole; blood incoagulable; a few ecchymoses in peritoneum.

The section of the cord was not quite complete.

Experiment No. 37.

	Time min. sec.	Pressure m. m.	REMARKS.
Normal	.	48	Injected intravenously 0.003 gram dried venom of the *Crotalus adamanteus* dissolved in 1 c. c. distilled water.
	10	46	
	20	48	
	30	46	
	1 00	48	
	1 20	46	
	3 20	40	
	3 50	32	
	5 50	36	
	7 50	30	Dead. Heart arrested in diastole; blood incoagulable; no ecchymoses in serous tissues.

Experiment No. 38.

	Time: min. sec.	Pressure m. m.	REMARKS.
Normal		30	Injected intravenously 0.003 gram dried venom of the *Crotalus adamanteus* dissolved in 1 c. c. distilled water.
	5	32	
	10	28	
	20	28	
	30	26	
	40	28	
	50	28	

Time. min. sec.	Pressure m. m.	REMARKS.
1 00	27	
1 30	27	
2 00	27	
2 30	22	
9 30	. .	Dead. Heart arrested in diastole; blood incoagulable, ecchymoses well-marked in peritoneum and pericardium; intestines congested; 50 c. c serum in peritoneal cavity. Section of cord complete, except anterior columns.

Experiment No. **39.**

	Time: min. sec.	Pressure m. m.	REMARKS.
Normal	. . .	44	Injected intravenously 0.015 gram dried venom of the *Crotalus horridus* dissolved in 1 c. c. distilled water.
	10	44	
	20	56	
	30	62	
	40	60	
	1 00	52	
	1 20	44	
	1 40	44	
	2 00	38	
	4 00	30	
	6 00	28	
	8 00	26	
	12 00	22	Dead.

Experiment No. **40.**

	Time min. sec.	Pressure m. m.	REMARKS.
Normal	. .	46	Injected intravenously 0.003 gram dried venom of the *Ancistrodon piscivorus* dissolved in 1 c. c. distilled water.
	20	40	
	30	48	Muscular movements.
	40	44	
	1 00	44	
	1 30	40	
	1 50	38	
	4 20	28	
	6 20	28	
	8 20	28	
	10 20	28	
	12 20	28	
	15 20	28	
	18 20	28	
	21 20	. . .	Dead. Blood is incoagulable; no ecchymoses in serous membranes.

Experiment No. **41.**

	Time: min. sec.	Pressure m. m.	REMARKS.
Normal	. . .	64	Injected intravenously 0.003 gram dried venom of the *Ancistrodon contortrix* dissolved in 1 c. c distilled water.
	10	56	
	20	54	
	40	48	

Time		Pressure	REMARKS
min.	sec.	m. m.	
1	00	42	
1	30	40	
2	00	42	
4	00	50	
7	00	56	
9	00	54	
11	00	48	
13	00	48	
15	00	48	
17	00	48	
20	00	48	
22	00	48	Injected 0.006 gram.
22	15	38	
22	30	32	Dead.

Experiment No. 42.

	Time	Pressure	REMARKS
	min. sec	m. m.	
Normal	. . .	56	Injected intravenously 0.003 gram dried venom of the Cobra
	10	60	dissolved in 1 c. c. distilled water and a few crystals of
	30	56	sodic chloride and filtered.
	1 00	52	
	3 00	48	
	5 00	48	
	8 00	52	
	11 00	58	
	14 00	60	
	19 00	62	Animal killed by pithing.

To recapitulate the actions of pure venoms upon the arterial pressure—we find that the injection of venom subcutaneously causes a progressive fall of blood pressure; when injected intravenously, there is a sudden and decided fall of pressure, which may be immediately followed by death, or by a gradual rise, to be in turn succeeded by a decline with feeble pulse as death approaches. In the Cobra there is a tendency to a rise of pressure, which may go above the normal as death appears.

After section of the pneumogastric nerves and its depressor fibres we find no alterations in the results obtained in normal animals, but when section of the cord is made in the middle or upper cervical region by which the vaso-motor centres in the medulla oblongata are practically destroyed, or when accompanying this section the nerves in the neck and the spinal cord in the middle cervical region are also cut, thus practically isolating the vaso-motor centres in the medulla and cutting off all central nervous connection with the heart, we find that the primary profound diminution of pressure is not so marked. There may even appear to be a slight tendency on the part of the arterial pressures to rise above the normal just before death. Even after section of the spinal cord, as above, we find in Cobra the increase of pressure occurring before death as in normal animals.

These results indicate that the primary positive failure of pressure is due chiefly to a depressant action of the venom upon the vaso-motor centres in the medulla oblongata, and slightly upon the heart. The tendency to a rise of pressure, as well as the ultimate fall, must be due to some action upon the heart itself or the general systemic capillaries. It seems probable that the rise of pressure in these experiments is of capillary origin since the pulse-curves do not indicate increased heart power, and we have already had reason to believe that venom exerts a decided action upon the capillaries themselves to bring about the remarkable ecchymoses found so commonly in cases of poisoning—an instance also of peripheral irritation, applicable here, is the effect of venom on the vagi peripheries in causing an increased respiration rate. The ultimate fall of pressure seems to be cardiac in origin, since there is an accompanying diminution in the force of the beats.

SECTION II.—THE ACTION OF VENOM GLOBULINS UPON THE BLOOD PRESSURE.

The Action of Venom Globulins upon the Blood Pressure of Normal Animals.— Thirteen experiments were made with the globulins upon normal animals. The doses usually given were those representing the amount of globulin in 0.015 gram of dried venom. The results of all of these experiments indicate that all the globulins exert an action analogous to that of the pure venom, but that they exhibit a material difference in the relative degree of their toxicity.

Of the thirteen experiments, seven were made with the *water-venom-globulin*, two with the *copper-venom-globulin*, and four with the *dialysis-venom-globulin*. Of the first series, five were made with the globulin from the *Crotalus adamanteus*; one with that of the *Ancistrodon piscivorus*, and one with that of the Cobra. The second and third series were made with globulins from the *Crotalus adamanteus* venom.

The *water-venom-globulin* produces the most profound changes, causing a primary diminution of pressure almost equalling that produced by pure venom, while *dialysis-venom-globulin* comes next; the *copper-venom-globulin* has but little effect. The actions of all of these globulins is to cause a primary fall of pressure, which is followed by a rise towards the normal and more or less well marked, while if the dose is sufficiently large the rise is followed by a fall to zero at death.

In one experiment made with the globulin from 0.035 gram of dried Cobra venom there was no appreciable effect. This was probably due to the very small proportion of globulin in this variety of venom.

Experiment No. 43.

	Time: min. sec.	Pressure m. m.	REMARKS
Normal	. . .	110	Injected intravenously 0.0012 gram *water-venom-globulin* (= 0.015 gram dried venom) from the dried venom of the *Crotalus adamanteus*.
	10	80	
	20	92	
	40	90	
	1 00	84	
	1 30	84	
	3 00	95	
	5 00	96	
	7 00	104	

THE ACTION OF VENOMS UPON ARTERIAL PRESSURE.

Time: min. sec.	Pressure m. m.	REMARKS.
9 00	106	
12 00	106	
15 00	110	
18 00	116	
25 00	124	
35 00	130	
45 00	130	
55 00	130	Animal killed by pithing. Heart in diastole; some ecchymoses in small intestine; blood remains fluid at the end of twenty-four hours—a few very soft clots are found.

Experiment No. 44.

	Time: min. sec.	Pressure m. m.	REMARKS.
Normal	. . .	130	Injected intravenously the *water-venom-globulin* from 0.03 gram dried venom of the *Crotalus adamanteus*.
	10	. . .	Pressure falling.
	20	96	
	40	96	
	1 00	94	
	1 20	90	
	1 40	90	
	3 40	94	
	5 40	102	Clot formed in the canula.
	7 40	108	
	9 40	108	Injected *water-venom-globulin* from 0.015 gram dried venom.
	10 00	104	
	10 20	104	
	10 40	104	
	14 00	106	
	17 00	106	
	20 00	108	
	30 00	102	Animal killed by pithing.

Experiment No. 45.

	Time: min. sec.	Pressure m. m.	REMARKS.
Normal	. . .	120	Injected intravenously 0.0033 gram *water-venom-globulin* (= 0.045 gram dried venom) from the dried venom of the *Crotalus adamanteus* dissolved by the addition of a trace of sodic carbonate.
	10	86	
	20	96	
	30	86	
	50	90	
	1 10	84	
	4 10	90	Injected 0.0066 gram as in the foregoing.
	4 30	80	
	4 35	104	
	4 40	100	Injected a similar dose.
	4 50	82	
	15	. . .	Killed by pithing. Heart arrested in diastole; few ecchymoses; blood remains fluid after twenty-four hours.

Experiment No. 46.

Time min. sec.	Pressure m. m.	REMARKS
Normal . . .	148	Injected intravenously the *water-venom-globulin* from one minim of fresh venom of the *Crotalus adamanteus*.
10	120	
20	116	
30	116	
50	100	
1 00	92	
1 30	82	
3 30	86	
5 30	96	
7 30	106	
9 30	118	
12 30	122	
14 30	126	
16 30	124	
17 30	100	
19 30	128	
21 30	132	Clot in canula.
26 00	128	
28 00	130	
30 00	124	
30 15	136	Clot in canula.
35 00	134	
37 00	136	
39 00	130	Animal killed by pithing; no ecchymoses; blood clots.

Experiment No. 47.

Time min. sec.	Pressure m. m.	REMARKS
Normal . . .	132	Injected intravenously the *water-venom-globulin* from 0.004 gram dried venom of the *Ancistrodon piscivorus* dissolved in 1 c. c. distilled water by the addition of a few crystals of sodic chloride.
30	126	
50	124	
1 00	136	
1 30	122	
1 50	114	Injected a similar dose.
2 00	106	
2 40	116	
3 10	104	

Experiment No. 48.

Time min. sec.	Pressure m. m.	REMARKS
Normal . . .	115	Injected intravenously the *water-venom-globulin* from 0.015 gram dried venom of the *Crotalus adamanteus*.
0	. . .	
10	. . .	
20	90	
30	93	
1 30	90	
2 30	100	
5 30	102	
10 30	100	
14 30	100	Hæmaturia.
19 30	103	

THE ACTION OF VENOMS UPON ARTERIAL PRESSURE.

Time: min. sec.	Pressure m. m.	REMARKS.
24 30	95	
29 30	85	
34 30	80	
42 30	75	
47 30	73	
52 30	60	
57 30	60	
67 30	57	
77 30	55	
80 30	38	
85	...	Dead; ecchymoses in intestines; blood incoagulable.

Experiment No. 49.

	Time: min. sec.	Pressure m. m.	REMARKS.
Normal	...	155	Injected intravenously *water-venom-globulin* from 0.035 gram dried Cobra venom dissolved in 1 c. c. distilled water.
	0	...	
	15	...	
	25	158	
	45	160	
	1 15	158	
	2 00	157	
	4 00	153	
	8 00	153	
	13 00	153	
	18 00	143	
	23 00	153	
	28	...	Broke loose from canula.

Experiment No. 50.

	Time: min. sec.	Pressure m. m.	REMARKS.
Normal	...	126	Injected intravenously **0.0012 gram** *copper-venom-globulin* (= 0.015 gram dried **venom**) from the dried venom of the *Crotalus adamanteus*.
	10	126	
	20	126	
	30	132	
	50	126	
	2 50	124	
	4 50	126	
	6 50	126	
	8 50	124	
	10 50	124	
	11 50	120	
	17 20	110	
	18 20	114	Injected double the foregoing dose.
	18 30	110	
	18 40	118	
	18 50	112	
	19 00	112	
	20 00	120	
	22 00	116	
	27	...	Killed. Heart in systole; few ecchymoses in lungs and intestines; blood remains fluid after two hours.

14 June, 1886.

Experiment No. 51.

Time: min. sec.	Pressure m. m.	REMARKS
Normal ...	112	Injected intravenously 0.0023 gram *copper-venom-globulin*
10	114	(= 0.03 gram dried venom) from the dried venom of the
30	110	*Crotalus adamanteus.*
1 00	112	
3 00	116	
5 00	120	
7 00	122	
8 00	122	Clot in canula.
10 00	124	
12 00	118	
22 00	118	
24 00	128	
26 00	124	Injected double the dose.
26 10	124	
26 30	116	
26 40	104	
27 00	108	
27 30	96	
29 30	99	
31 30	98	
34 30	104	
39 00	116	
41 00	116	
43 00	116	
45 00	118	
52 00	120	
58 00	122	Killed by pithing; no ecchymoses.

Experiment No. 52.

Time: min. sec.	Pressure m. m.	REMARKS
Normal ...	132	Injected intravenously 0.0017 gram *dialysis-venom-globulin*
20	124	from the dried venom of the *Crotalus adamanteus* dissolved
40	112	in 1 c. c. distilled water with a trace of sodic carbonate.
50	116	
1 20	108	
3 20	108	
5 20	120	
18 20	130	
18 23	...	Injected 0.0034 gram *dialysis-venom-globulin.*
18 30	96	
18 45	100	
19 05	94	
19 25	102	
19 55	96	
20 25	76	
20 55	60	
21 25	46	
22 00	42	
30 00	...	Dead. No ecchymoses; heart in diastole; blood remains fluid at the end of **one** hour.

Experiment No. 53.

Time: min. sec.	Pressure m. m.	REMARKS
Normal . . .	126	Injected intravenously *dialysis-venom-globulin* from the dried
20	120	venom of the *Crotalus adamanteus* (quantity unknown).
30	114	
1 00	110	
1 20	102	
1 40	102	
2 00	100	
2 40	100	
3 40	102	
5 40	110	
6 20	114	
6 50	118	
7 20	124	
7 50	126	
8 50	128	
9 20	128	
9 50	130	
10 50	134	
11 50	122	
12 50	112	
14 20	112	Injected more of the *globulin*.
14 50	84	
15 20	78	
15 50	62	Killed by pithing.

Experiment No. 54.

Time: min. sec.	Pressure m. m.	REMARKS
Normal . . .	150	Injected intravenously 0.0017 gram *dialysis-venom-globulin*
10	118	from the dried venom of the *Crotalus adamanteus*.
20	130	
30	118	
1 00	116	
2 00	110	
5 00	126	
7 00	124	
16 00	136	Injected 0.0034 gram.
17 30	116	" "
17 40	100	
18 30	104	
20 30	126	
23 30	118	
28 30	102	
43 30	98	
53 30	94	Animal killed.

The Action of Venom Globulins upon the Blood Pressure of Animals in which the Pneumogastric Nerves had been Severed.—Four experiments were made on animals in which the pneumogastric nerves and depressor nerves were severed. The results in these experiments do not differ in quality from those obtained in normal animals; the effects, however, appear to be less decided than in animals with the pneumogastrics intact. Here, as in the previous experiments, the copper-venom-globulin exhibits comparatively little effect on the pressure.

Of the four experiments which were made with globulins from the *Crotalus adamanteus*, one was made with the *water-venom-globulin*; two with the *copper-venom-globulin*, and one with the *dialysis-venom-globulin*. It will be noticed that in several instances considerable rises of pressure occurred accompanied by struggles; the former effect being, no doubt, due to the latter, and not to a peculiar action of the globulin.

Experiment No. 55.

	Time: min. sec.	Pressure m. m.	REMARKS.
Normal	. . .	116	Injected intravenously 0.0011 gram *water-venom-globulin* from
	10	100	the dried venom of the *Crotalus adamanteus*.
	20	100	
	30	110	
	1 00	106	
	1 40	110	
	3 40	110	
	5 40	108	Clot in canula. Animal killed by pithing.

Experiment No. 56.

	Time: min. sec.	Pressure m. m.	REMARKS.
Normal	. . .	130	Injected intravenously 0.0024 gram *copper-venom-globulin*
	10	132	from the dried venom of the *Crotalus adamanteus*.
	20	132	
	30	132	
	40	132	
	1 10	132	
	3 10	132	
	5 10	128	
	7 10	128	
	9 10	128	
	23 10	136	Injected a similar dose.
	23 20	130	
	23 40	132	" " "
	23 55	132	
	24 00	116	Respiration greatly slowed.
	24 15	116	Animal broke loose. Killed by pithing.

Experiment No. 57.

	Time: min. sec.	Pressure m. m.	REMARKS.
Normal	. . .	116	Injected intravenously 0.0012 gram *copper-venom-globulin*
	20	116	from the dried venom of the *Crotalus adamanteus*.
	40	116	
	50	116	
	2 50	116	
	4 50	116	
	6 50	116	
	8 50	112	

Time: min. sec.	Pressure m. m.	REMARKS.
11 50	116	
13 50	118	
15 50	118	Injected a double quantity.
16 10	100	
16 20	118	" " "
16 30	110	
16 45	108	
17 45	132	Struggles.
19 45	114	
21 45	112	
23 45	106	
25 45	100	
26 45	88	
27 00	...	Animal killed by pithing.

Experiment No. 58.

Time: min. sec.	Pressure m. m.	REMARKS.
Normal	120	Injected intravenously 0.0017 gram *dialysis-venom-globulin*
10	100	from the dried venom of the *Crotalus adamanteus*.
20	112	
30	110	
50	150	Struggles.
1 50	190	"
4 20	130	
6 20	120	
8 20	126	
10 20	122	
12 20	120	
17 50	118	
18 20	118	Injected 0.0034 gram.
18 30	100	
18 40	148	Struggles. Injected a similar dose.
19 00	140	"
19 15	170	"
19 20	176	"
21 50	144	
22 00	140	
23 00	166	
25 00	122	
27 00	128	
29 00	114	Struggles.
34 00	82	
34 30	80	
38 30	60	
41 00	50	
47 00	28	
49 00	...	Dead.

The Action of Venom Globulins upon the Blood Pressure of Animals in which the Pneumogastric, Depressor, and Sympathetic Nerves and Cervical Spinal Cord had been Cut.—Four experiments were made on animals in which the nerves of the neck and the cord in the middle or upper cervical region (excepting one) were cut. They were all made with the globulins from the *Crotalus adamanteus*; one with *water-venom-globulin*, one with *copper-venom-globulin*, and two with *dialysis-venom-globulin*.

The results of this series of experiments accord with those observed when pure venom was used, and with the preceding experiments with the globulins. The primary fall of pressure is slight, while the tendency to a secondary rise is very marked, since in three of the experiments the pressure rose above the normal. The action of water-venom-globulin on the primary fall was most marked, while in the single experiment made with copper-venom-globulin, in which eight times the quantity was given in two doses, the pressure rose slightly, and continued above normal. When the dose is sufficient to kill, the pressure ultimately gradually declines, accompanied by a feeble pulse.

In the last experiment with *dialysis-venom-globulin* it will be noticed that tremors are accompanied with a rise of pressure during their existence.

Experiment No. 59.

	Time: min. sec.	Pressure m. m.	REMARKS.
Normal	. . .	48	Section of cord made below the 6th cervical vertebra. Injected intravenously 0.0011 gram *water-venom-globulin* from the dried venom of the *Crotalus adamanteus*.
	10	38	
	30	38	Artificial respiration stopped.
	1 00	48	
	1 10	34	
	2 10	46	
	3 10	42	
	5 10	40	
	7 10	58	
	9 10	34	
	11 10	30	
	15 10	30	
	17 40	30	
	19 00	30	
	21 00	30	
	23 00	30	
	25 00	30	
	27 00	30	Animal killed by pithing.

Experiment No. 60.

	Time: min. sec.	Pressure m. m.	REMARKS.
Normal	. . .	29	Injected intravenously 0.0048 gram *copper-venom-globulin* from the dried venom of the *Crotalus adamanteus*.
	10	32	
	20	30	
	40	32	
	1 00	33	
	3 30	38	
	5 30	40	
	7 30	42	

THE ACTION OF VENOMS UPON ARTERIAL PRESSURE. 111

Time: min. sec.	Pressure m. m.	REMARKS.
9 30	42	
11 30	42	
13 30	40	
16 30	44	
17 00	42	Injected a similar quantity.
17 30	38	
18 00	40	
20 00	44	
24 00	46	
26 00	44	
28 00	42	
30 00	40	
32 00	40	
34 00	40	Animal killed by pithing.

Experiment No. 61.

Time: min. sec.	Pressure m. m.	REMARKS.
Normal ...	42	Injected intravenously 0.0017 gram *dialysis-venom-globulin* from the dried venom of the *Crotalus adamanteus*.
10	40	
30	40	
1 00	46	
3 00	44	
6 00	40	
8 30	38	
10 30	38	
12 30	...	Injected 0.0068 gram.
12 50	44	
13 10	48	
13 30	46	
13 50	48	
15 00	36	
16 00	46	
18 00	28	Dead. **Heart** arrested in diastole; no ecchymoses; blood fluid. **A few fibres** of the anterior columns of the cord were uncut.

Experiment No. 62.

Time: min. sec.	Pressure m. m.	REMARKS.
Normal ...	44	Injected intravenously 0.0068 gram *dialysis-venom-globulin* from the dried venom of the *Crotalus adamanteus*.
10	40	
20	40	
30	44	
40	52	Universal tremors persistent.
50	62	
1 00	62	
1 10	60	
1 20	58	Clot in canula.
3 20	86	Blood pressure fell very low before this observation, and was raised by the tremors returning vigorously.
3 50	52	
6 50	22	
8 50	8	
9 00	...	Dead. No ecchymoses; blood incoagulable; heart natural.

From this series of experiments with globulins it seems clear that they possess the peculiar physiological effects of pure venoms upon the blood pressure; that the water-venom-globulin is the most powerful, and the copper-venom-globulin the least so, and that the copper-venom-globulin seems to exhibit a more marked tendency than the others to cause a rise of pressure.

SECTION III.—THE ACTION OF VENOM PEPTONES UPON THE BLOOD PRESSURE.

The Action of Venom Peptones upon the Blood Pressure of Normal Animals.— Seven experiments were made with the peptones from different venoms: two with that of *Crotalus adamanteus*; three with *Ancistrodon piscivorus*; and two with *Cobra*. The action of peptones upon the blood pressure is similar to that observed with the pure venom and the globulins, but their power to cause the primary profound fall of pressure is certainly much less, while the rise of pressure after the primary fall is decidedly more marked, and there is also a tendency to go above the normal. In two experiments, one with the peptone of the Crotalus and one with that of the Moccasin, the pressure was not primarily reduced, but there was a rise above the normal from the first. Where the animal was watched until death the pressure was observed to undergo a more or less gradual decline with feeble heart-beats. In several instances a rise of pressure was noted which was usually due to convulsive seizures.

Experiment No. 63.

	Time: min. sec.	Pressure m. m.	REMARKS.
Normal	. . .	140	Injected intravenously the *peptone* from 0.015 gram dried venom of the *Crotalus adamanteus*.
	10	128	
	20	128	
	30	136	
	40	128	
	50	128	
	1 00	124	
	11 00	116	
	21 00	116	Clot.
	49	. . .	Dead. No ecchymoses; lungs seem congested; blood clots readily.

Experiment No. 64.

	Time: min. sec.	Pressure m. m.	REMARKS.
Normal	. . .	114	Injected intravenously the *peptone* from 0.03 gram dried venom of the *Crotalus adamanteus*.
	10	130	
	30	132	
	1 00	120	
	2 00	140	
	5 00	144	
	9 00	124	Killed. Ecchymoses in the lungs; blood clots.

THE ACTION OF VENOMS UPON ARTERIAL PRESSURE.

Experiment No. 65.

Time: min. sec.	Pressure m. m.	REMARKS.
Normal . . .	86	Injected intravenously the *peptone* from 0.015 gram dried venom of the *Ancistrodon piscivorus*.
30	88	
1 00	88	
1 30	88	
4 30	92	
10 30	94	Injected double the amount.
10 50	100	
11 10	102	
11 30	96	

Experiment No. 66.

Time: min. sec.	Pressure m. m.	REMARKS.
Normal . . .	116	Injected intravenously the *peptone* from 0.015 gram dried venom of the *Ancistrodon piscivorus*.
10	94	
20	66	
30	70	
1 00	76	
1 30	74	
2 00	76	
2 30	76	
3 00	76	
5 00	66	Injected a similar quantity.
5 20	60	
5 40	66	
6 00	68	
6 30	66	
7 00	62	Injected a double quantity.
7 30	56	
8 00	60	

Experiment No. 67.

Time: min. sec.	Pressure m. m.	REMARKS.
Normal . . .	140	Injected intravenously the *peptone* from 0.05 gram dried venom of the *Ancistrodon piscivorus*.
10	94	
20	100	Convulsions.
30	160	"
40	170	"
50	190	"
1 00	130	"
1 20	130	
1 40	142	
2 00	136	
2 20	136	
2 40	136	
5 40	118	Injected 0.005 *peptone*.
5 50	114	
6 00	104	
6 20	112	
6 30	112	
9 30	118	
13 30	122	Killed.

15 June, 1886.

Experiment No. 68.

Time: min. sec.	Pressure m. m.	REMARKS
Normal ...	140	Injected intravenously the *peptone* from 0.005 gram dried Cobra venom.
10	130	
20	148	
30	140	
40	142	
1 00	142	
1 20	140	
3 20	138	
6 20	152	
11 20	224	Convulsive twitchings.
16 20	176	
16 50	134	
17 50	92	
18 20	104	Blood is asphyxiated; no respiration.
18 50	84	
19 00	46	
19 10	46	
19 30	38	
19 50	36	
20 10	40	
20 30	49	Killed.

Experiment No. 69.

Time: min. sec.	Pressure m. m.	REMARKS
Normal ...	130	Injected intravenously the *peptone* from 0.06 gram dried Cobra venom.
10	130	
15	116	
30	156	
50	156	
3 20	140	
8 20	190	Tonic convulsions.
10 20	176	Convulsive twitchings; asphyxiated blood; respiration ceased.
10 30	208	
10 40	182	
10 50	136	
11 00	116	
11 20	102	
11 40	86	
12 00	66	
12 20	54	
12 40	38	Dead.

The Action of Venom Peptones on the Blood Pressure of Animals with Pneumogastric and Depressor Nerves Severed.—After section of the pneumogastric and depressor nerves the results are not appreciably altered. Four experiments were made: one with *Crotalus adamanteus*; one with *Ancistrodon piscivorus*, and two with *Cobra* venom. In all of these experiments the pressure during the secondary rise went above the normal.

THE ACTION OF VENOMS UPON ARTERIAL PRESSURE. 115

Experiment No. 70.

Time: min. sec.		Pressure m. m.	REMARKS.
Normal	. . .	160	Injected intravenously the *peptone* from 0.015 gram dried venom of the *Crotalus adamanteus*.
	10	160	
	15	140	
	30	150	
	40	156	Struggles.
	50	156	
1	00	134	
1	20	158	
1	30	164	
3	30	126	
8	30	122	Struggles.
5	30	146	
20	30	148	
25	30	140	
30	30	140	
35	00	136	
40	00	148	
45	00	142	
50	00	140	
62	00	138	Killed.

Experiment No. 71.

Time: min. sec.		Pressure m. m.	REMARKS.
Normal	. . .	124	Injected intravenously the *peptone* from 0.015 gram dried venom of the *Ancistrodon piscivorus*.
	10	100	
	20	140	
	30	150	
	40	130	
	50	140	
1	00	134	
1	20	114	
1	40	116	
2	00	136	
2	30	124	
2	40	110	
2	50	124	
3	00	136	
3	10	110	
3	20	96	
4	20	. . .	Respiration ceased; heart beats.

Experiment No. 72.

Time: min. sec.		Pressure m. m.	REMARKS.
Normal	. . .	138	Injected intravenously the *peptone* from 0.005 gram dried Cobra venom.
	10	142	
	20	136	
	40	134	
1	00	140	
1	30	138	

Time : min. sec.	Pressure m. m.	REMARKS.
3 30	136	
5 30	136	
7 30	132	
9 30	132	
10 00	140	
21 30	146	Twitchings.
25 30	142	
26 30	150	Killed. No ecchymoses.

Experiment No. 73.

Time : min. sec.	Pressure m. m.	REMARKS.
Normal ...	124	Injected intravenously the *peptone* from 0.006 gram dried Cobra venom.
10	122	
20	128	
40	124	
1 00	126	
3 00	120	
5 00	120	
16 00	118	
31 00	...	Dead. Asphyxiated; no ecchymoses; blood clots in canula.

The Action of Venom Peptones on the Blood Pressure of Animals in which the Pneumogastric, Depressor, and Sympathetic Nerves and Cervical Spinal Cord were Cut.—In five experiments on animals in which the nerves in the neck and the spinal cord in the middle or upper cervical region were cut we found that but little alteration occurred in the blood pressure until late in the poisoning, excepting in one experiment with the *Ancistrodon piscivorus*, in which the pressure sank immediately and death occurred in thirty seconds. Two experiments were made with the peptone of the *Crotalus adamanteus*; one with the *Ancistrodon piscivorus*, and two with the Cobra. In all of these experiments, excepting one with Cobra, there was an immediate comparatively slight fall of pressure after injection, which was followed generally by a rise; in the excepted case of the Cobra there was a primary rise equal to 3 m. m. of mercury, which was followed by a fall, and this in turn by a rise. The pressure, as in the previous experiments, usually declines towards death.

Experiment No. 74.

Time : min. sec.	Pressure m. m.	REMARKS.
Normal ...	50	Injected intravenously the *peptone* from 0.015 gram dried venom of the *Crotalus adamanteus*.
10	50	
20	48	
40	48	
1 00	48	
3 00	44	
6 00	42	
11 00	42	
13 00	42	
15 00	42	

THE ACTION OF VENOMS UPON ARTERIAL PRESSURE.

Time: min. sec.	Pressure m. m.	REMARKS.
17 00	50	Injected the *peptone* from 0.03 gram dried venom.
17 30	50	
19 30	48	
21 30	48	
23 30	46	
25 30	46	
27 30	44	
29 30	44	
31 30	42	
34 30.	38	Dead. No ecchymoses; blood fluid after fifteen minutes.

Experiment No. 75.

	Time: min. sec.	Pressure m. m.	REMARKS.
Normal	. . .	67	Injected intravenously the *peptone* from 0.015 gram dried venom of the *Crotalus adamanteus*.
	15	50	
	20	50	
	30	50	
	1 00	50	
	1 30	50	
	2 00	50	

Experiment No. 76.

	Time: min. sec.	Pressure m. m.	REMARKS.
Normal	. . .	50	Injected intravenously the *peptone* from 0.015 gram dried venom of the *Ancistrodon piscivorus*.
	10	45	
	18	38	
	30	33	
	1 00		Dead.

Experiment No. 77.

	Time: min. sec.	Pressure m. m.	REMARKS.
Normal	. . .	128	Injected intravenously the *peptone* from 0.01 gram dried Cobra venom.
	10	122	
	20	132	
	30	134	
	1 00	132	
	12 00	138	
	12 30	136	
	17 30	. . .	Dead. Blood clots readily. Ecchymoses in base of lungs.

Experiment No. 78.

Time: min. sec.	Pressure m. m.	REMARKS.
Normal ...	38	Injected intravenously the *peptone* from 0.015 gram dried Cobra venom.
11	39	
20	41	
40	40	
1 40	35	
5 40	35	
10 40	37	
15 40	38	
16 00	...	} Injected a similar dose.
16 06	38	
16 15	38	
16 25	43	
16 45	43	
17 00	43	
18 00	40	
20 00	25	
23 00	...	Dead.

From all of the results of these experiments it seems justifiable to conclude that the isolated principles of venoms exert the poisonous actions of pure venoms on the blood pressure, and that their toxic effects are essentially simply different in degree. These various poisons all play a part in the alterations of pressure, acting towards the same end, but mainly with different degrees of intensity; the *water-venom-globulin* appears to be the most potent in the pressure alterations, the *dialysis-venom-globulin* next, then the *peptone*, and finally the *copper-venom-globulin*. The globulins are the more active in the production of the diminution of pressure, and the peptone in the secondary rise.

The globulins no doubt play a very important part in the poisonous phenomena of Crotalus poisoning, a less important part in *Ancistrodon* poisoning, and but very little in Cobra poisoning; these differences not depending as much upon differences in the quality of the globulins in the species of venom to which they belong as on differences in quantity.

CHAPTER IX.

THE ACTION OF VENOMS AND THEIR ISOLATED GLOBULINS AND PEPTONES UPON RESPIRATION.

Section I.—Pure Venom.

In our experiments on respiration rabbits were always used, and the rate of breathing was recorded on a revolving drum by the lever of a Marey's tambour, the latter being connected with the animal by means of a tracheal tube. The injections in all of the experiments, excepting two, which were subcutaneous, were made into the external jugular vein.

In experiments on normal animals we observed no qualitative difference in the several venoms used. Ten experiments were made upon normal animals: four with the venom of the *Crotalus adamanteus*; three with that of the Moccasin, *piscicorus*, and three with that of the Cobra. In eight of these experiments there was a primary increase in the respiration rate followed by a diminution far below the normal, while in two the respirations were at once diminished, and became persistently slower until death. In both of these cases death occurred very soon after injection, indicating a most profound action of the poison.

Action of the Pure Venoms on the Respirations in Normal Animals.

Experiment No. 1.

	Time: min. sec.	Respirations per minute.	Length of curve m. m.	REMARKS.
Normal	. . .	84	10	Injected intravenously 0.002 gram dried venom of the
	10	180	16	*Crotalus adamanteus* dissolved in 1 c. c. distilled
	40	84	12	water.
	1	96	. . .	
	2 20	108	. . .	
	4 50	72	8	
	6 50	72	6	
	8 50	60	6	
	10 50	48	. . .	Convulsive movements.
	11 20	40	4	" "
	11 50	24	4	" "
	12 20	26	. . .	
	12 50	. . .	. . .	Dead.

Experiment No. 2.

Time: min. sec.	Respirations per minute.	Length of curve m. m.	REMARKS.
Normal ...	42	6	Injected intravenously 0.004 gram dried venom of the *Crotalus adamanteus* dissolved in 1 c. c. distilled water.
10	43	10	
40	?	...	Struggles, which prevent a count.
1 00	84	12	
2 10	30	25	Convulsive movements.
3 40	9	23	Conjunctival reflexes gone.
5 00	10	14	
5 10	...	...	Respiration ceased. Heart still beating. The respiratory muscles respond to stimulus. The spinal cord was exposed, and the motor columns were found to respond to electrical stimulus. The motor nerves responded after the motor columns of the cord had lost their irritability.

Experiment No. 3.

Time: min. sec.	Respirations per minute.	Length of curve m. m.	REMARKS.
Normal ...	84	7	Injected intravenously 0.006 gram dried venom of the *Crotalus adamanteus* dissolved in 3 minims distilled water.
20	158	3	
40	120	7	
1 00	96	11	
1 30	90	10	
2 00	96	12	
2 30	102	10	
3 00	120	7	Struggles.
3 30	35	5	
4 30	60	6	
5 30	35	10	Conjunctival reflexes gone.
6 30	10	7	
7 30	4	4	Respiration ceased. Respiratory muscles irritable. The spinal cord was quickly exposed; the sensory columns give no response, the motor columns are active. The motor columns of the cord fail before the motor nerves.

Experiment No. 4.

Time: min. sec.	Respirations per minute.	Length of curve m. m.	REMARKS.
Normal ...	66	6	Injected intravenously 0.015 gram dried venom of the *Crotalus adamanteus* dissolved in 1 c. c. distilled water.
15	...	...	Arrest of respiration attended with a tetanic condition.
30	36	16	
1 00	12	16	
1 50	18	6	
2 20	?	5	
2 25	...	...	Respiration ceased. Spinal cord rapidly exposed and tested by electrical currents; sensory columns fail first, then the motor columns, then motor nerves.

THE ACTION OF VENOMS UPON RESPIRATION.   121

Experiment No. 5.

Time: min. sec.	Respirations per minute.	Length of curve m. m.	REMARKS
Normal ...	100	9	Injected intravenously 0.004 gram dried venom of the *Ancistrodon piscivorus* dissolved in 5 minims distilled water.
10	210	8	
20	150	21	
30	140	20	
40	120	23	
50	...	...	Convulsions.
1 10	...	...	Dead.

Experiment No. 6.

Time: min. sec.	Respirations per minute.	Length of curve m. m.	REMARKS
Normal ...	135	6	Injected intravenously 0.004 gram dried venom of the *Ancistrodon piscivorus* dissolved in 1 c. c. distilled water.
10	420	...	Struggles. Respiration at once began to increase rapidly, and reached a maximum rapidity during the occurrence of struggles.
20	270	6	
30	65	18	Tetanic movements.
40	60	...	" "
50	120	10	
1 00	120	...	" "
1 10	60	...	
1 20	90	12	
6 20	60	...	
11 20	180	...	
16 20	210	...	
16 30	...	...	Killed.

Experiment No. 7.

Time: min. sec.	Respirations per minute.	Length of curve m. m.	REMARKS
Normal ...	144	6	Injected intravenously 0.004 gram dried venom of the *Ancistrodon piscivorus* dissolved in 1 c. c. distilled water.
10	300	8	
20	240	12	
30	150	11	
5 00	60	10	
7 00	80	7	
12 00	54	7	
18 00	70	7	
18 05	...	...	Injected as above 0.008 gram venom.
18 30	210	9	
18 40	160	10	
18 50	80	5	
23 50	65	7	Killed by pithing.

122   THE VENOMS OF CERTAIN THANATOPHIDEÆ.

Experiment No. 8.

Time: min. sec.	Respirations per minute.	Length of curve m. m.	REMARKS.
Normal . . .	60	9	Injected intravenously 0.015 gram dried Cobra venom
20	80	. . .	dissolved in 1 c. c. distilled water and filtered.
40	120	15	Struggles.
1 00	100	45	"
1 20	60	10	
1 40	42	38	
2 00	. . .	. . .	Respiration ceased.

Experiment No. 9.

Time: min. sec.	Respirations per minute.	Length of curve m. m.	REMARKS.
Normal . . .	300	8	Injected intravenously 0.015 gram dried Cobra venom
10	300	9	dissolved in 1 c. c. distilled water.
20	255	11	
30	285	10	
40	240	10	
1 00	255	11	
1 30	200	9	
2 00	150	7	
2 10	125	7	
3 20	. . .	. . .	Respiration ceased.

Experiment No. 10.

Time: min. sec.	Respirations per minute.	REMARKS.
Normal . . .	36	Injected intravenously 0. 03 gram dried Cobra venom in solution.
20	39	
40	39	
1 00	51	
1 30	52	
2 00	45	
4 00	48	
9 00	46	
12 00	36	
14	. . .	Respiration ceased.

The Action of Pure Venoms on the Respiration in Animals in which the Pneumogastric Nerves were Cut.—When injections are made, after section of the pneumogastric nerves, the primary increase in the respiration rate does not occur, but a diminution begins at once; and, on the whole, drops irregularly until death ensues. Four experiments were thus made: one with *Crotalus adamanteus*, two with *Ancistrodon piscivorus*, and one with Cobra venom; the results being on the whole reasonably uniform.

THE ACTION OF VENOMS UPON RESPIRATION. 123

Experiment No. 11.

Time: min. sec.	Respirations per minute.	Length of curve m. m.	REMARKS.
Normal ...	42	7	Injected intravenously 0.002 gram dried venom of the *Crotalus adamanteus* dissolved in 1 c. c. distilled water.
30	20	18	Slight struggles preceding this observation interfered with the marker.
1 00	28	10	
1 30	6	22	
2 00	6	19	
2 30	3	25	
3 00	12	18	
3 30	6	21	
4 30	6	21	
5 30	6	15	
6 30	8	12	

Experiment No. 12.

Time: min. sec.	Respirations per minute.	Length of curve m. m.	REMARKS.
Normal ...	103	15	Injected intravenously 0.004 gram dried venom of the *Ancistrodon piscivorus* dissolved in 1 c. c. distilled water.
10	94	15	
20	84	30	Struggles.
30	102	25	
40	77	25	
50	60	28	
1 00	60	25	
1 10	45	30	
1 15	...	...	Dead.— Respiration ceased; heart still beats. Animal dies in tetanus.

Experiment No. 13.

Time: min. sec.	Respirations per minute.	REMARKS.
Normal ...	102	Injected intravenously 0.003 gram dried Cobra venom dissolved in 1 c. c. distilled water.
30	57	
1 00	78	
2 00	57	
4 00	66	
10	...	

Experiment No. 14.

Time: min. sec.	Respirations per minute.	Length of curve m. m.	REMARKS.
Normal ...	127	17	Injected intravenously 0.004 gram dried venom of the *Ancistrodon piscivorus* dissolved in 1 c. c. distilled water.
20	92	47	
30	90	32	
40	82		
50	82		
1 05	67	..	
1 30	...	...	Respiration ceased.

In none of these experiments do we find a primary increase in the respiration rate, as in animals with intact vagi, but invariably a diminution. It seems clear, therefore, that the first result must be dependent upon an excitation of the peripheries of the pneumogastric nerves, and that the diminution of respirations is due to a centrally active cause. Should the lessened number of the respirations be central, that is, dependent upon a depression of the respiratory centres, we would expect to find that the degree of depression would depend upon the relative amount of venom coming in contact with these centres in a given space of time. We have accordingly made an experiment, in which this suggestion is admirably carried out by injecting the venom into the carotid artery, thus throwing the poison directly upon the respiratory centres.

Experiment No. 15.

	Time: min. sec.	Respirations per minute.	Length of curve m. m.	REMARKS.
Normal	...	78	42	Injected into the right carotid artery 0.015 gram of
	15	7	38	dried venom of *Crotalus adamanteus* dissolved in 1
	30	4	30	c. c. distilled water.
	1 00	4	25	
	1 30	10	45	Convulsions.
	2 00	...	...	Dead.

It seems obvious from the preceding experiments that venoms exert a double action on the respiration; first, an irritant action on the peripheries of the pneumogastric nerves, by which an increase in the respiration rate is brought about; and secondly, a depression of the respiratory centres, by which the respiration rate is diminished. Since the diminution in the respirations occurs in animals with cut pneumogastrics immediately after injection, and at a time when an increase occurs in normal animals, it is apparent that these two factors are acting in normal animals at the same time to produce opposite results; consequently, whether we have an increase or a decrease in the respirations must be dependent upon the relative degree of power exerted by one or the other of these factors. In most cases we have found a primary increase of respirations followed by a diminution; it is therefore obvious that the action of the venom upon the peripheries of the pneumogastric nerves was more than able to compensate for the depressant action of the poison upon the respiratory centres; this is very clear since no increase of respirations above normal occurs in animals with cut pneumogastrics. In the two cases in normal animals in which a decline from the first was observed, and in which the animals died in a few minutes after injection, the action of the venom upon the respiratory centres was so profound that the accelerator factor was unable to cause a rise. This is also illustrated in the experiment in which the venom was injected into the carotid artery and thrown upon the respiratory centres.

Since venom does not seem to exert other than a depressant action upon the respiratory centres, it does not appear probable that it would have an opposite effect upon the respiratory nerves, so that the effect of the venom upon the peripheries of the pneumogastric nerves is probably one of irritation rather than stimulation, and probably due to some secondary cause, which is likely to be located in the profound alteration of the blood or the destructive action of the venom upon the pulmonary tissues, as illustrated, for instance, upon capillaries.

SECTION II.—THE ACTION OF GLOBULINS ON THE RESPIRATIONS.

The Action of Venom Globulins upon the Respiration in Normal Animals.— Seven experiments were made with globulins upon normal animals: three with the *water-venom-globulin* of the *Crotalus adamanteus;* and one with the *water-venom-globulin* of Cobra; one with the *copper-venom-globulin*, and one with *dialysis-venom-globulin*, both from the *Crotalus adamanteus*.

These poisons, excepting the *copper-venom-globulin*, all act like the pure venoms, but generally with a less degree of intensity, causing a primary acceleration of the respiration followed by a decline. In the second experiment, however, there was no diminution, but the respirations became enormously increased so that at death they were nearly trebled in frequency. The *copper-venom-globulin* does not cause any primary acceleration, but simply a diminution.

Experiment No. 16.

	Time: min. sec.	Respirations per minute.	Length of curve m. m.	REMARKS.
Normal	. . .	100	9	Injected intravenously the *water-venom-globulin* from
	20	100	15	0.015 gram dried venom of the *Crotalus adamanteus*.
	40	100	12	
	1 00	96	11	
	2 00	96	9	
	4 00	120	10	
	5 00	120	10	
	6 00	132	10	
	8 00	108	10	
	10 00	90	9	
	12 00	80	9	
	14 00	69	7	
	14 05	. . .	. . .	Injected as above from 0.06 gram dried venom.
	14 20	80	15	
	14 40	60	10	Struggles.
	16 40	90	10	
	18 40	96	8	
	20 40	108	9	
	22 40	114	9	
	24 40	108	9	
	26 40	90	6	
	28 40	96	8	
	30 40	72	7	
	32 40	64	8	
	34 40	70	8	
	36 40	75	8	
	38 40	80	9	
	40 40	90	10	
	44 00	. . .	. . .	Dead. Heart arrested in diastole; some ecchymoses.

Experiment No. 17.

Time: min. sec.	Respirations per minute.	Length of curve m. m.	REMARKS.
Normal	75	8	Injected intravenously 0.0158 gram *water-venom-globulin* (5 days old) from the dried venom of *Crotalus adamanteus*.
15	80	8	Struggles.
30	60	...	
1 00	60	...	
1 30	90	...	
6 30	110	...	
11 30	110	...	
16 30	110	...	Injected the same as above.
17 00	120	...	
27 00	130	...	
29 00	...	...	Dead. Blood remains fluid; some ecchymoses.

Experiment No. 18.

Time: min. sec.	Respirations per minute.	REMARKS.
Normal	114	Injected intravenously the *water-venom-globulin* from 0.035 gram of dried Cobra venom dissolved in 1 c. c. distilled water.
15	...	
25	126	
45	132	
1 15	150	
2 00	150	
6 00	204	
11 00	114	
16 00	84	
26 00	62	
33 00	60	
56 00	63	
66 00	62	
76 00	72	
120	...	Killed. Animal in fair condition.

Experiment No. 19.

Time: min. sec.	Respirations per minute.	REMARKS.
Normal	73	Injected intravenously the *water-venom-globulin* from 0.015 gram dried venom of the *Crotalus adamanteus*.
20	84	
30	78	
1 30	120	
2 30	108	
5 30	96	
10 30	96	
14 30	82	Hæmaturia.
19 30	70	
24 30	75	
29 30	78	
34 30	72	
42 30	69	
47 30	72	
52 30	96	

THE ACTION OF VENOMS UPON RESPIRATION. 127

Time: min. sec.	Respirations per minute.	REMARKS.
57 30	90	
67 30	84	
77 30	84	
80 30	12	
85	...	Dead. Ecchymoses generally; **blood fluid**.

Experiment No. 20.

Time: min. sec.	Respirations per minute.	Length of curve m. m.	REMARKS.
Normal ..	180	20	Injected intravenously the *copper-venom-globulin* from
20	174	19	0.015 gram dried venom of the *Crotalus adamanteus*.
40	168	19	
1 00	168	19	
3 30	168	20	
5 30	108	15	
7 30	100	9	
9 30	110	10	
11 00	168	17	
13 00	138	11	
15 00	120	10	
17 00	144	10	
19 00	102	9	
21 00	108	10	
23 00	104	7	
25 00	112	9	
27 00	100	9	
30 00	90	7	
34 00	90	7	
39 00	112	10	
41 00	85	8	
43 00	96	7	
45 00	108	5	
47 00	108	7	
49 00	76	6	
51 00	66	7	
53 00	80	8	
57 00	96	8	
59 00	90	9	
60 00	116	10	
63 00	118	8	
65 00	116	10	Struggles.
69 00	104	9	
71 00	100	7	
73 00	116	7	
75 00	100	8	
77 00	116	11	
79 00	140	11	
81 00	130	10	
85 00	130	10	
87 00	120	10	
91 00	126	9	
92 00	...	...	Killed by pithing. Lungs very much ecchymosed; abdominal viscera normal; heart normal; blood coagulates.

Experiment No. 21.

Time min. sec.	Respirations per minute.	Length of curve m. m.	REMARKS
Normal . . .	54	18	Injected intravenously 0.0012 gram *water-venom-*
10	60	16	*globulin* from the dried venom of the *Crotalus ada-*
26	54	15	*manteus.*
40	54	15	
1 00	48	16	
3 00	60	17–42	Struggles.
5 00	72	33	
7 00	54	30	
8 00	63	32	
10 00	60	28	
11 30	60	30	
13 30	72	42	
15 30	70	38	
17 00	78	45	Injected 0.0022 gram *water-venom-globulin.*
17 10	78	42	
17 20	102	38	
18 20	72	38–78	Struggles.
19 20	66	22	
21 20	60	23	
24 20	70	25	
27 20	60	29	Injected 0.0024 gram *water-venom-globulin.*
27 40	60	26	
28 20	60	25	
29 50	. . .	. . .	Killed by pithing; some ecchymoses.

Experiment No. 22.

Time min. sec.	Respirations per minute.	Length of curve m. m.	REMARKS
Normal . . .	112	9	Injected intravenously the *dialysis-venom-globulin* from
10	120	12	0.015 gram dried venom of the *Crotalus adamanteus.*
20	160	16	
30	140	16	
40	140	15	
1 00	140	16	
2 00	126	10	
5 00	156	14	
10 00	174	15	
12 00	130	14	
13 30	142	10	Injected *dialysis-venom-globulin* from 0.06 gram of
13 50	150	22	dried venom.
14 30	132	13	
19 00	130	9	
24 00	120	8	
29 00	110	10	
39 00	80	6	
54 00	80	3	
55 00	. . .	. . .	Dead. Respiration ceased before the heart. Ecchymoses in the lungs and in the pericardium, in the small intestine, ureters, and bladder.

THE ACTION OF VENOMS UPON RESPIRATION.

The Action of Venom Globulins on the Respiration of Animals in which the Pneumogastric Nerves were Cut.—Two experiments were made on animals with cut pneumogastric nerves: one with the *dialysis-venom-globulin*, and one with the *copper-venom-globulin*, both from the *Crotalus adamanteus*.

In neither experiment was there an increase in the respirations; these results being in accord with the experiments made with pure venom.

Experiment No. 23.

Time: min. sec.	Respirations per minute.	Length of curve m. m.	REMARKS.
Normal . . .	42	8	Pneumogastric nerves previously cut. Injected hypo-
10	39	10	dermically the *dialysis-venom-globulin* from 0.015
20	30	12	gram dried venom of the *Crotalus adamanteus*.
40	24	8	
1 00	27	10	
1 20	?	. . .	Struggles.
3 20	20	9	
5 20	35	8	
8 20	24	5	
18 20	30	5	
23 20	32	6	
36 20	42	7	Injected *dialysis-venom-globulin* from 0.06 gram.
36 40	42	7	
38 40	24	20	Struggles.
42 00	. . .	. . .	Respiration ceased; heart beats feebly; blood remains incoagulable; great ecchymoses in abdominal viscera.

Experiment No. 24.

Time: min. sec.	Respirations per minute.	Length of curve m. m.	REMARKS.
Normal . . .	60	5	Pneumogastric nerves previously cut. Injected intra-
30	54	6	venously the *copper-venom-globulin* from 0.015 gram
1 00	48	5	dried venom of the *Crotalus adamanteus*.
3 00	48	5	
8 00	48	5	
12 00	52	6	
15 00	48	6	Injected *copper-venom-globulin* from 0.03 gram dried
15 30	30	4	venom in two doses.
15 40	54	15	Struggles with very irregular breathing followed by
16 00	?	. . .	gasping respiration.
16 30	18	10	
19 00	12	4	
24 00	20	4	
27 00	20	5	
30 00	30	5	
35 00	26	5	
39 00	27	5	
41 00	30	5	
44 00	42	6	
49 00	30	6	Injected *copper-venom-globulin* from 0.12 gram dried venom in two doses.
54 00	. . .	. . .	Respiration ceased; heart still beats; ecchymoses in heart and lungs marked.

17 June, 1886.

130   THE VENOMS OF CERTAIN THANATOPHIDEÆ.

The results of these experiments with the globulins indicate that the *water-venom-globulin* and *dialysis-venom-globulin* act like the pure venom, while the *copper-venom-globulin* lacks the property of producing the primary acceleration of the respirations.

SECTION III.—THE ACTION OF VENOM PEPTONES ON THE RESPIRATION.

The Action of Venom Peptones on the Respiration in Normal Animals.—Three experiments were made on the normal animals with the venom *peptones*; in two with the peptone from the *Crotalus adamanteus*, and in one with the peptone from the *Ancistrodon piscivorus*. In all of these experiments the increase of the respiration rate was strongly marked.

Experiment No. 25.

	Time: min. sec.	Respirations per minute.	Length of curve m. m.	REMARKS.
Normal	. . .	225	68	Injected intravenously the *peptone* from 0.03 gram
	10	255	60	dried venom of the *Crotalus adamanteus* obtained
	30	255	60	by boiling.
	1 00	300	56	
	2 00	270	50	
	5 00	270	50	
	9 00	270	55	Killed. Blood clots readily; moderate ecchymoses in the lungs.

Experiment No. 26.

	Time: min. sec.	Respirations per minute.	Length of curve m. m.	REMARKS.
Normal	. . .	180	11	Injected intravenously the *peptone* from 0.06 gram
	10	240	16	dried venom of the *Ancistrodon piscivorus* obtained
	30	270	15	by boiling.
	40	240	. . .	
	1 00	345	. . .	
	1 10	270	. . .	
	1 20	240	. . .	
	4 20	240	. . .	
	9 20	300	. . .	
	18 20	360	14	
	28 20	270	. .	
	38 20	180	11	

THE ACTION OF VENOMS UPON RESPIRATION. 131

Experiment No. 27.

Time: min. sec.	Respirations per minute.	Length of curve m. m.	REMARKS.
Normal . . .	75	9	Injected intravenously the *peptone* from 0.015 gram
10	120	42	dried venom of the *Crotalus adamanteus*.
3 00	30	10	
6 00	75	8	
11 00	50	8	
18 00	48	7	
23 00	50	7	
28 00	45	7	
33 00	60	7	
37 00	60	9	
49	. . .	. . .	Dead. No ecchymoses; lungs slightly congested.

In one animal the increase was equal to one-third of the normal; in the second, in which a larger dose was used, the normal rate was doubled; and in the third it rose to more than one-half of the normal. There was not, however, in any of the animals that marked depression which is observed in poisoning with pure venom or venom globulins.

The Action of Venom Peptones on the Respiration in Animals in which the Pneumogastric Nerves had been previously Divided.—In one experiment in which the pneumogastric nerves were cut and in which the peptone from the venom of the *Crotalus adamanteus* was used, the well-marked primary increase in the respirations did not occur, there being a diminution from the first.

Experiment No. 28.

Time: min. sec.	Respirations per minute.	Length of curve m. m.	REMARKS.
Normal . . .	80	13	Injected intravenously the *peptone* from 0.015 gram
10	52	18	dried venom of the *Crotalus adamanteus*.
15	37	12	
20	25	8	
30	22	7	
1 00	30	8	
1 30	30	7	
3 30	40	9	
8 30	60	8	Struggles.
13 30	36	13	
20 30	48	17	
25 30	50	14	
30 30	52	15	
35 00	55	12	
40 00	48	15	
45 00	45	15	
50 00	45	16	
62 00	44	15	Killed by pithing.

In this experiment, as in those with pure venom and venom globulins in which the animals had the pneumogastrics cut, the increased respiration rate seen in normal animals did not occur.

The results of the experiments with venom peptone are therefore in accord with those with the pure venom and the venom globulins.

Summary.—From the results of the observations with pure venoms and their globulins and peptones upon the respiration it seems clear that the primary action of all of the above poisons, excepting the copper-venom-globulin, is to cause an increase in the number of respirations, and secondarily to diminish the respirations below the normal. Of the different principles the peptone seems to exert the most decided power in causing the acceleration, while the copper-venom-globulin seems to utterly lack this action.

Since the primary increase of the respirations does not occur in any case after section of the pneumogastric nerves, this effect must be exerted by an action of the poisons upon the peripheries of these nerves, and since after section of these nerves a diminution of the respirations always occurs this effect must be due to a depression of the respiratory centres, as we have found that the motor nerves and muscles of respiration are irritable long after the cessation of this function.

CHAPTER X.

PATHOLOGY.

Pathology of Serpent Venoms.—The pathology of snake poisoning in man owes most of what is best in our knowledge of it to the researches of the East Indian surgeons and to American observers.

In the following observations Prof. H. F. Formad has followed with great success the lines of a research which were laid down with care by the authors of this essay. They have also been at great pains to repeat, and to verify, most of the observations made by this distinguished observer.

The Nature and Character of the Individual Morphological Constituents of Venom.—Having seen that fresh venom consists morphologically of a liquid and of a solid part, it was necessary to ascertain the exact nature and character of each.

The following means were resorted to:—

1st. The separation of the granular material (of fresh venom) by filtration and the submission to physiological tests of the liquid filtrate and of the solid residue, each separately.

2d. The exposure of fresh venom to a temperature high enough to kill organized life, and then submitting it to physiological tests.

3d. Studying the effects of venom and of its isolated morphological constituents upon dead animal substances. (Putrefaction and other experiments.)

4th. The isolation and culture of the organisms contained in venom and the testing of the physiological effects of these isolated and washed organisms (viz., of pure cultures of micrococci).

1st. *Filtration Experiments with Fresh Venom.*—On account of its viscid and glutinous character venom could not be satisfactorily filtered except under a high pressure through a vacuum filter. About two drachms of fresh *Crotalus adamanteus* venom were forced by means of a hydraulic air pump through a porous clay cylinder such as is employed in certain small galvanic batteries, or else the venom was filtered through a thin layer of plaster of Paris moulded in the neck of a small glass filter. The liquid filtrate obtained was perfectly clear, and examined under the microscope showed no organic or solid particles of any kind. The solid residue left upon the filter consisted of granular material, such as has been described before, of bacteria and a few cells. This residue was diligently and repeatedly washed with boiled distilled water, by passing the latter through the filter.

The amount of residue (about three grains) just obtained was dried and introduced subcutaneously into the pectoral muscle of a pigeon, but without effect.

Two pigeons were injected in the pectoral muscle, one with five, and the other with two minims of the liquid filtrate above described, and both died promptly within six minutes and twenty-five minutes respectively.

2d. *Experiments with Heated Venom.*—Fresh Crotalus venom rapidly dried was put in a covered watch-glass and subjected for one hour to a temperature of 115° C. in the dry-heat oven. The venom was thereby converted into a dense resinous opaque brown mass.

Two grains of this mass, upon the addition of distilled water forming a turbid liquid, were divided into thirds and injected hypodermatically into a rabbit, a rat, and a pigeon, respectively. The rabbit died in 15 minutes, the rat in 12 minutes, and the pigeon in 7 minutes, after the operation, with results and lesion similar to those obtained by the use of fresh venom.

This experiment also shows that the virulence of venom does not reside in any of its organized constituents.

3d. *Putrefaction Experiments.*—The testing of the effects of venom on various dead animal substances was particularly desirable on account of the remarkable capacity of the venom to induce rapid putrefaction in the tissues of living animals. It was necessary to learn whether this property of bringing about speedy necrotic changes was an action inherent in venom or due to any of its accidental constituents.

Putrefaction Experiments with Sterilized Bouillon and Fresh Venom and its Active Principles (not Sterilized).—This bouillon was prepared from chicken in the same manner as that ordinarily used for culture liquids for bacteria, and the experiments were executed in a room at a temperature of about 70° F.

About two drachms of sterilized bouillon were put in each of sixteen ordinary test tubes which were then treated as follows :—

Tubes 1 and 2, added to bouillon one drop of fresh Crotalus venom ; mouth of tubes plugged with cotton.

Tubes 3 and 4, prepared same as last, but tube left open (no cotton plug).

Tubes 5 and 6, added one grain of Crotalus peptone. Tube closed by cotton plug.

Tubes 7 and 8, same as last, but tubes left open.

Tubes 9 and 10, added one grain of Crotalus globulin. Tubes closed.

Tubes 11 and 12, same as last. Tubes open.

Tubes 13 and 14, a pure bouillon, nothing added to it. Tubes closed.

Tubes 15 and 16, same as last. Tubes open.

Twenty-four hours later the bouillon in all the test tubes which originally was perfectly clear had become cloudy except tubes 13 and 14 (which contained the sterilized pure bouillon plugged well with cotton).

On the third day of the experiment tubes 3 and 4 (fresh venom, tubes open) showed well-pronounced putrefaction of the bouillon.

Slight putrefactive changes were subsequently observed in the remaining tubes (except 13 and 14) in the following order :—

On the fourth day, tubes 7 and 8. On the fifth day, tubes 11 and 12, also in tubes 15 and 16.

On the seventh day all the plugged specimens were examined, and all showed more or less putrescence except the tubes with the pure bouillon as stated. Of these closed test tubes, however, tubes 1 and 2 (the fresh venom) showed the putrefactive changes to be much more pronounced than in the remaining tubes; but as we have seen putrefaction ensued much sooner in the tubes that were open (tubes 3 and 4).

As all the tubes showed putrefaction more or less, it is presumable that the *peptone* and *globulin* accidentally contained bacteria, these substances not having been sterilized at the commencement of the experiment.

The contents of the tubes examined microscopically during and at the end of the experiment showed the presence of bacteria of putrefaction in direct proportion to the putrefaction changes.

Imperfect as this experiment may be, it appears to establish the fact that fresh venom promotes putrefactive changes comparatively more rapidly than the venom peptone and globulin, but it also shows further that this power to produce putrescence is very much aided by the action of the air, and depends upon the presence of bacteria contained in that air or in the venom. It was also evident that putrefaction was considerably retarded in all the tubes that were plugged by the cotton, and further that unplugged tubes containing sterilized soup, and exposed to contamination from air showed also putrefaction but at a later date.

Putrefaction Experiment with Muscular Tissue and Venom.—The following rough experiment also appears to show that putrefactive changes develop in dead animal tissues much more rapidly in the presence of venom than without it.

Experiment.—A few drops of a solution of dry Crotalus venom were poured upon a small piece of fresh muscle just removed from the thigh of a rabbit and placed in a covered glass beaker.

A similar preparation but without the addition of venom was made in a second covered beaker. Temp. 70° to 80° F.

Putrefactive changes began to appear in the specimen treated by the venom after twenty-four hours, and after seventy-two hours were quite far advanced. Under the microscope the muscular tissue showed necrotic alterations very similar to those (to be described later) as occurring in experiments upon the living muscle. A multitude of dumb-bell-shaped rod bacteria, some large bacilli and the micrococci of the venom enormously multiplied, were seen in the decaying muscular substance.

In the specimen of muscle not treated by venom, putrefactive changes were delayed to the fifth day and then appeared to be much less conspicuous, showing but few bacteria. The muscle fibres were uniformly cloudy and degenerated but not broken down in the peculiar manner caused by venom.

Experiments with Bouillon and Venom in Sealed Glass Bulbs, Venom being thoroughly Sterilized.—More satisfactory and conclusive results were obtained from the following experiments:—

A number of small glass bulbs were filled with sterilized bouillon after the well-known method of Dr. Sternberg, and after being thoroughly resterilized by boiling the following preparations were made:—

To each of six bulbs was added one grain of dry Crotalus venom, the venom having been previously subjected to sterilization in a dry heat at 110° C. for one hour. The bulbs were then hermetically sealed by melted glass. The bouillon in these tubes (with sterilized venom) remained perfectly clear and free from bacteria. Microscopical examination was made at various periods, the last time after eighteen months when it was still perfectly clear and showed no signs of putrefaction.

A similar result was obtained in an experiment with another set of six glass bulbs filled with bouillon, and to which some Moccasin peptone, previously sterilized, was added. These bulbs looked somewhat cloudy, but on examination of the contents eighteen months later no bacteria, and no putrefactive changes were noted.

As a control experiment six bulbs filled with pure sterilized bouillon were kept for a similarly long period, and they all remained clear and free from change; while a few bulbs filled with unsterilized bouillon showed great cloudiness, bacteria, and putrefactive change.

4th. *Culture Experiments.*—The study of the morphology of the bacteria inhabiting the venom was next undertaken. To this end numerous culture experiments to isolate the bacteria from the venom were made. As stated before, the perfectly fresh venom contained only one form of these vegetable organisms, the micrococci, and only to these latter attention was paid; the rod bacteria and bacilli not appearing except in venom which had begun to putrefy.

The micrococci contained in the venom showed the following behavior in pure cultures: Of culture soils, the peptonized gelatine prepared after the formula of Koch proved to be quite suitable. The isolation of the micrococci was made after the methods of Sternberg and of Koch, as adopted in the pathological laboratory of the University of Pennsylvania. For gelatine culture a minute quantity of venom was smeared on the surface of the solidified jelly contained in a sterilized, small, flat, well covered glass vessel. The micrococci liquefied the jelly, an effect not peculiar to all bacteria. After twenty-four hours all over the inoculated surface of the jelly were seen small turbid drops which contained the micrococci. With a sterilized platinum wire the micrococci from one of the liquefying specks upon the first culture were transplanted to the jelly in a second culture vessel. From this second generation a minute quantity was transplanted to a third and fourth culture vessel. The fourth and all the later generations yielded usually a pure crop of micrococci.

In impure cultures dumb-bell-shaped bacteria and sometimes large bacilli were met with. These, however, could not be said to be peculiar to venom, as they are never found in fresh venom. It may, therefore, be concluded that these cultures represent the micrococci peculiar to, or at least those constantly inhabiting venom. Moreover, the micrococci in these cultures whenever they were successful, were the only bacterium seen and were fully identical as to shape, **measurement,** and behavior to aniline dyes with those found in the fresh venom.

Much better crops of the venom-micrococci were obtained in bouillon cultures in Sternberg's glass-bulbs. The micrococci grow more rapidly and better in these bulbs, because **the bouillon can be heated** up to the more suitable temperature of 40° C.; while the jelly cultures could not be warmed to such a degree without melting solid gelatine. A variety of other culture soils and methods of isolation were employed in these experiments, but their description here is unnecessary as not being sufficiently related to the points at issue.

In relation to the morphology of the micrococci it may be added, that they measure on the average $\frac{1}{10000}$ of an inch in diameter; they often appear in pairs, but most commonly in zooglœa masses. They show a distinct aureole, such as is met with in various forms of micrococci.[1]

These aureoles have lately been erroneously described by Friedländer, as peculiar to certain "specific" micrococci in croupous pneumonia. In conclusion, it might be said that the venom micrococci do not appear to differ from the micrococci found in the saliva of men and other animals.

In order to test whether the venom-micrococci were in any way specific or pathogenetic, and whether they form, or contribute to, the virulence of the venom, inoculations with pure cultures of the micrococci were made upon animals.

As these experiments gave entirely negative results, it is superfluous to enter into details. Suffice it to say that large quantities of the pure micrococci from a sixth generation were injected, in various manners, into rabbits, cats, pigeons, and white rats, but without fatal results; or without producing any other lesion than occasionally local abscesses, or later on, metastatic abscesses. Sometimes the so-called "miliary tuberculosis of animals" was produced by inoculating with the venom-micrococci. No signs of any lesions resembling those of venom poisoning were observed.

Experiments made to Study the Anatomical Changes produced by the Venom in living Animals. Naked Eye Appearances.—Very many years ago Dr. Weir Mitchell described two forms of venom poisoning—*rapid or acute*, and *slow or chronic*. To the latter appear to be relegated by him all those cases in which death is protracted beyond a few hours. This convenient division is justified by certain differences in the mode of termination of venom poisoning, and by the macroscopic and microscopic appearances of the lesions induced.

In the most rapid poisoning, there is frequently nothing appreciable to the naked eye beyond the slight local lesion or here and there minute capillary hemorrhages, when death has been delayed beyond a minute. In examples of chronic poisoning both the local and the systemic changes are enormously more extensive. When animals were subjected to chronic poisoning they were kept under the influence of narcotics, since it had been learned that these agents did not affect the results. No Cobra venom was employed in this series, but only the pure or dried venoms of our own serpents, or else some one or other of the constituents of these poisons.

The following tables relate the experiments made, and the more striking morphological changes:—

---

[1] See "Memoir on Diphtheria," Report to the National Board of Health, 1882, by H. C. Wood and H. F. Formad.

138   THE VENOMS OF CERTAIN THANATOPHIDEÆ.

RAPID POISONING.

Effects of Venom when Injected Hypodermatically into or Applied otherwise to the Tissues of the Living Animal.

No. of expt.	Animal used.	Form and quantity of venom, and where introduced.	Time of death.	Local lesion.	Condition of blood.	Changes in thorax, abdomen, brain, and membranes.	REMARKS.
1	Pigeon	Crotalus venom, fresh, ¼ grain injected into pectoral muscle	Killed after 5 minutes	Moderately sized, dark hemorrhagic swelling	Coagulable and red	All internal organs congested; no other changes visible; no ecchymoses perceptible	This animal was killed before the full effects of the venom.
2	Pigeon	Same as last	Died in 15 minutes	Very dark colored hemorrhagic swelling	Less coagulable and quite dark	Organs only moderately congested, but there were numerous small subdural, subperitoneal, and slight subpericardial ecchymoses	For the details and the histological appearances, see the next chapter. The studies of the changes in muscular tissue were mostly made from this experiment.
3	Pigeon	Moccasin venom, fresh, 1 drop injected into peritoneum	Died after 1 hour and 50 minutes	Profuse hemorrhage all over peritoneal cavity	Liquid very dark	Ecchymoses in nearly all organs, quite marked in arachnoid and at base of brain; some so small as to be visible only by microscope. Extreme congestion	
4	Pigeon	Crotalus venom, fresh, 1 drop into peritoneum	Died in 25 minutes	Same as last	Liquid dark	Hemorrhages only subperitoneal, other organs merely congested	
5	Rabbit	Moccasin venom, injected into peritoneum	? minutes	Subperitoneal hemorrhages	Coagulable on exposure	No changes beyond local lesion	
6	Pigeon	Peptone, injected into pectoral muscle	35 minutes	Hemorrhagic swelling	Liquid	No visible changes, except all organs congested	Changes in muscular tissue similar to those produced by fresh venom, but far less blood effused.
7	Pigeon	Moccasin venom, fresh, injected into cavity of skull	5 minutes	Hemorrhages in arachnoid and brain tissue	Slightly coagulable	Membranes of brain and brain substance peripherally soaked with blood; other organs congested	
8	Rabbit	Moccasin venom, fresh, ½ grain injected into lung	1 minute	Lung infarcted by blood	Same as last	No changes, except in lung and some subpericardial ecchymoses	See specimen and description in chapter on histological changes.
9	Rabbit	Peptone, ½ grain into liver and peritoneum	43 minutes	Ecchymosis locally only	Liquid dark	Other organs not visibly affected	
10	Rabbit	Same as last	Killed at the end of 1 hour	Slight subperitoneal ecchymoses	Dark, but coagulable	No systemic changes.	
11	Pigeon	Peptone, 1 grain into peritoneum	20 minutes	Local ecchymosis slight	Ditto	No notable changes in other organs	Histological changes similar to those produced by fresh venom.
12	Pigeon	Globulin, ½ grain into peritoneum	40 minutes	The same as last	Ditto		
13	Cat	Crotalus globulin, ½ grain into peritoneum	40 minutes	Profuse ecchymosis	Slightly coagulable	Hemorrhage only local	
14	Cat	Crotalus peptone, ½ grain into peritoneum	1 hour and 20 minutes	Same as last, but less marked	Liquid	Hemorrhage only local, also extreme congestion of all organs	Microscopic examination made of every organ. The details will be given hereafter.
15	Rabbit	Crotalus globulin, ½ grain into peritoneum	Killed in 10 minutes	Same as last	Red and coagulable	Nothing peculiar beyond the local lesion	
16	Rabbit	Dry Crotalus venom, ½ grain in watery solution, into peritoneum	Killed in 10 minutes	Same as last	Same as last	Same as last	
17	Rabbit	Dry Crotalus venom, ½ grain in watery solution, into peritoneum	Died 1 hour and 25 minutes	Same as last	Very dark, liquid	Ecchymoses in all organs examined. Profuse ecchymoses in peritoneum, also subpleural, subarachnoid, and subpericardial. Liver, which was injured by the syringe, showed a large hemorrhagic infarction	Some of the ecchymoses were so small as to be visible only on microscopical examination.

RAPID POISONING —Continued.

No. of expt.	Animal used.	Form and quantity of venom, and where introduced.	Time of death.	Local lesion.	Condition of blood.	Changes in thorax, abdomen, brain, and membranes.	Remarks.
18	Cat	Dry Crotalus venom, 1 grain in watery solution, into peritoneal cavity	Died 5¼ hours	Same as last	Same as last	Peritoneal cavity contains a good deal of liquid blood; hemorrhage at base of brain (subarachnoid); no other lesion noted; organs rather anæmic. Heart empty, contracted	Cats appear to resist the effects of venom much longer than the other animals used in this research.
19	Cat (chloralized)	Peritoneum opened, mesentery exposed uninjured, in moist chamber, and smeared repeatedly with a solution of dry venom, using not less than 5 grains of venom	Died after 4 hours and 35 minutes	Hemorrhagic infiltration, quite extensive, but came on very slowly	Partly coagulable	Peritoneal hemorrhage; organs anæmic	It appears that when the mesentery is exposed and not injured the animal survives much larger applications of venom than if venom be injected into an unopened peritoneal cavity. Very small quantities of venom appear to kill in the latter case. For further experiments of this character, see Mechanism of Hemorrhages.

None of the cases in the table exhibit instances of the greatest possible rapidity of death. Dr. Mitchell has seen a pigeon die within ten seconds from a hypodermatic injection of pure Crotalus venom. In such a case there is positively no lesion, and the blood is solidly coagulated.

In most cases very soon after injection of the venom in either of its forms, the time varying from a few minutes to a few hours, according to the kind of animal and the quantity of venom used, there appears a swelling at the point of injection with intense violet-black discoloration of the skin, which gradually extends over an area of several square inches. On making an incision into the tissues in the immediate neighborhood of the injection, they are found to be soaked with extravasated blood. This is often all that is visible if death has occurred soon; but if it has been postponed for a short time, then in tissues distant from the place of the injection, extravasations to a smaller extent were often found. Most pronounced and most frequent are the ecchymoses below serous membranes (subpleural, subperitoneal, and subpericardial); in fact the whole organism is deeply affected, the tissues being congested and presenting a much darker appearance than normal. The blood does not seem to coagulate readily within cavities or interstices of the body *unless death follows almost instantaneously*. In cases which live longer, the blood remains commonly in a liquid state, or coagulates imperfectly, and then only after being exposed to the air, resembling in this particular the state of that fluid observed in conditions of asphyxia.

SLOW POISONING.

Effects of Venom upon the Tissues of the Living Animal.

No. of expt.	Animal used.	Form and quantity of venom, and where introduced.	Time of death.	Local lesion.	Condition of blood.	Changes in internal organs.	REMARKS.
20	Pigeon	Copper globulin, 2 c. c. (equal to 1 gram fresh venom) injected into pectoral muscle	15 hours	Large, dark gangrene like swelling of chest; muscle disintegrated	Liquid and dark	Subpericardial ecchymoses and pericardial effusion. Red tinged serum in peritoneal cavity. Heart empty. Lungs and pleura full of ecchymoses. All the organs congested	
21	Rabbit	Unknown, but very minute quantity of Crotalus venom injected in back	9 days	Dark gangrenous swelling	Liquid and dark	Numerous minute hemorrhages below serous membranes, seen also at base of brain in right posterior horn. The organs rather anæmic and softened	All these autopsies were made immediately or quite shortly after death.
22	White rat	Crotalus venom, dry, ½ grain injected into abdomen	2 days and 7 hours	Hemorrhagic peritonitis	Slightly coagulable, dark	Organs congested, softened; nothing else peculiar found; small, loose, red clot in right side of heart	For changes in blood, see details in text.
23	Cat	Crotalus venom, dry, 1 grain injected into right thigh	9 days and 2 hours	Skin slough over local lesion, which is dark, hemorrhagic, and gangrenous	Liquid dark, ill smelling	All internal organs softened and highly ecchymosed and congested. Fæces and urine bloody. Hemorrhage at base of brain, and minute blood specks in pericardium. Heart quite atrophied and softened	
24	Pigeon	Quantity unknown, injected into pectoral muscle	14 days	Atrophy, with pigmentation of the pectoral muscle injected	Liquid and very dark	Hemorrhages indicated by deposits of blood pigment in the tissues. All the organs in a state of atrophy and softened, resembling acute yellow atrophy in man. Serous sacks all distended with bloody serum. Heart empty, and although contracted quite soft	For further details of the histological changes, see text. N. B.—Gangrenous changes in the local lesion are usually more pronounced in the "Slow" than in the Rapid form of venom poisoning.

The following lesions may be mentioned as peculiar to retarded or slow poisoning: Rigor mortis often quite absent. The blood, usually diffluent, is very dark and does not readily acquire the scarlet-red color when exposed to the air. There are prominent blood-stained effusions in all the serous sacks. (Plate V.) Urine and fæces often bloody. Hemorrhages beyond the local lesion much more conspicuous than in the rapid poisoning. The remote lesions of slow poisoning resemble very much (morphologically) the primary local lesion, but are not so extensive or so well defined. In general the conditions of slow venom poisoning resemble those of acute septic poisoning. It is very often impossible to draw a distinct line between the manifestations of rapid and slow poisoning, nevertheless the division is in practice convenient.

One case of very protracted slow poisoning was observed in a pigeon which had been injected with venom in the pectoral muscle. (See Experiment 24, Table Slow Poisoning.)

Instead of the usual gangrenous change there was seen in this case after the lapse of two weeks a decided dry atrophy of the muscular tissue about the wound. Its fibres were greatly diminished in size as compared with the opposite unaffected muscle, and many of them were entirely disintegrated, as was evident from the remnants of the muscular fibres and the granular material which took their place between the interstices of the connective tissue. This granular material was seen throughout the specimen, some of it being of a brown tint, and probably representing disintegrated blood corpuscles. The internal organs were all in a state of atrophy, more particularly so the liver, the tissues of which under the microscope bore a striking resemblance to acute yellow atrophy. The serous sacks were all largely distended by blood stained serum. The heart muscle was also in a condition of atrophy, its chambers empty, and the blood dark and not coagulable. Blood examined microscopically showed appearances to be mentioned shortly.

The Effects of certain Venoms on the Coagulability of the Blood.—One of the most interesting differences in the action of the venoms of the Rattlesnake and Cobra and which was pointed out some years ago by more than one observer, is that the former venom partially or completely destroys the coagulability of the blood, while the venom of the Cobra has no such marked effect. The blood of animals poisoned with Crotalus venom is usually thin and dark, the clots form slowly, and are very soft and easily broken up.

Some direct studies were made to test more accurately this interesting property of the Crotalus venom, and it was thus observed that it is not peculiar to the poison of this genus, but is also a characteristic of the Mocensin. Several of these observations which were made with the venom of the *Crotalus adamanteus* we record in detail.

Experiment.—Five test-tubes were used :—

No. 1 empty.
No. 2 contained $\frac{1}{4}$ grain dried venom dissolved in 0.5 c. c. distilled water.
No. 3 " $\frac{1}{4}$ " " in 1.0 " " "
No. 4 " $\frac{1}{2}$ " " in 1.0 " " "
No. 5 " 2 drops glycerine solution of venom, equal parts.

These test-tubes were packed in snow, to retard coagulation and to give the venom a better opportunity to act, the tubes remaining in this condition for about half an hour. The main artery in the leg of a large etherized rooster was exposed, and a canula placed in it. The blood was allowed to flow into the tubes in the order of their numbers, the tubes being gently shaken to mix the venom and blood. The operation began at 3:55 and ended at 4:00 P. M. At 4:35 the tubes were examined. The blood in No. 1, which contained no venom, was firmly clotted, in all the others the blood was fluid. At 4:55 the test-tubes were all taken from the snow. Blood in No. 1 was firmly clotted and of a bright-red color; blood in Nos. 2, 3, and 4 was fluid and venous in appearance; blood in No. 5 was fluid and of a brighter red than No. 1. The tubes were then corked with raw cotton and set aside.

Twenty-four hours later blood in No. 1 was firmly clotted, in Nos. 2 and 3 tarry, in No. 4 tarry, but thinner than in Nos. 2 and 3, in No. 5 perfectly fluid; in the lower half of the tube was a mass of corpuscles, while the upper half had the appearance of pure serum.

Forty-eight hours—no appreciable alteration.

Seventy-two hours—no appreciable alteration; the blood in tube No. 1 had no unpleasant odor, but all the rest gave decided odors of putrefaction, and were very dark.

Comparative observations were also made at the same time with different venoms, using as before a fowl to furnish us the blood and the snow pack to retard coagulation.

In test-tube No. 1 was placed ½ grain dried Moccasin venom in 1 c. c. distilled water.

In test-tube No. 2 was placed ½ grain dried Moccasin venom boiled and filtered through clay.

In test-tube No. 3 was placed ½ grain of dried Crotalus venom in 1 c. c. distilled water.

In test-tube No. 4 was placed ½ grain of dried Crotalus venom in 1 c. c. distilled water, heated gradually to 70° C.

In test-tube No. 5 was placed ½ grain dried Cobra venom in 1 c. c. distilled water.

In test-tube No. 6 was placed ½ grain dried Cobra venom in 1 c. c. distilled water, boiled and filtered through a clay filter.

In test-tube No. 7 nothing was placed but the pure blood.

Into each of the test-tubes about 10 c. c. of blood was allowed to flow; at the end of 15 minutes the blood in Nos. 2, 5, 6, and 7 was clotted firmly, and the blood in Nos. 1, 3, and 4 was perfectly fluid. After one hour and a quarter the blood in No. 1 was clotted in a quite remarkable clot, which was exceedingly elastic—the clot when picked up and suspended drew out into a long worm-like thread, and could then be further pulled out to at least double its length, resuming its natural size when placed upon the table. The blood in No. 3 had some very soft clots. In No. 4, the blood was clotted soft.

On the second day all of the bloods were firmly clotted except Nos. 1, 3, and 4, which were perfectly fluid and had a putrefactive odor, which was absent in the others. On the third day these bloods were clotted and had some dark serum, but the pure blood was clotted firmly and perfectly dry on the surface.

From these observations it seems clear that the Cobra venom exerts no appreciable effect on the coagulability of the blood of a chicken when thus circumstanced, and that Crotalus and Moccasin venoms act powerfully. Moreover, that the effect of the Crotalus venom is the more efficient, and that if the solutions of venom have been subjected to a degree of heat sufficient to coagulate the venom-globulins, the effect is lessened very greatly. It thus appears that the principle affecting the coagulability of the blood is most largely the globulin.

It would seem therefore that venom-peptone, although not without power to lessen the coagulability of blood, has not the full efficiency of the globulins. Neither can it be said as to this capacity, that the small percentage of Cobra globulin has even relatively the anti-clotting capacity of the globulins of Crotalus.

These differences between Cobra and Crotalus show themselves strikingly in the slighter local disorders caused by the Indian serpent.

The singular formation of an elastic clot was observed in other cases. It appeared to be a temporary condition, and to be in a measure due to the great increase in the adhesiveness of the blood corpuscles.

Microscopical Changes in the Various Tissues of the Body from the Effects of the Venom of Crotalus.

Effects of Fresh Venom upon the Blood Corpuscles.—A series of experiments were made to study the direct as well as the remote effects of the venom upon the blood corpuscles. The result of these observations was the discovery of some changes which have not been heretofore fully described.

A drop of blood from man or any mammal treated with a minute quantity of fresh venom, presented the following appearances under the microscope: Upon the white blood corpuscles, the venom did not appear to have any other effect than to stop the amœboid motion, which in presence of venom could not be kept up, even by the use of the warm stage. The cells appeared somewhat larger than usual and also more granular. The red blood corpuscles appeared unchanged when observed, but for a moment, and superficially, yet prolonged and careful study revealed very remarkable alterations. The alterations in the red blood corpuscles are essentially these:—

The blood disks lose their biconcavity and assume a spherical form, but without parting with their coloring matter. They exhibit also great adhesiveness, arranging themselves into various sized and shaped aggregations. The corpuscles comprising these groups sometimes appear to fuse so that their outlines cannot be determined, even by high amplification. In addition the corpuscles seem to soften and acquire a peculiar ductility and capacity to be stretched without fracture. By inclining the stage of the microscope, or making gentle pressure upon the cover-glass, allowing thereby the liquid to flow, the red blood corpuscles may be seen to elongate themselves into spindle-shaped or even into fine thread-like bodies. (Figs. 1 and 2, Plate III., and Plate IV.) Such masses of corpuscles appear to act like colloid material.

One drop of human blood was mingled with one of fresh snake venom by the application of the cover-glass. The three fields photographed were found in the zones of contact between the blood and venom; they occurred within a small area—almost adjacent. Fields 1, 2, and 3, Plate IV., were photographed respectively within 15, 30, and 40 minutes after the first application of the venom to the blood. As the masses of corpuscles were slowly changing form and position, the exposure was, necessarily, but for a part of a second. The lens employed was Spencer $\frac{1}{8}$ immersion, giving 400 diam. with the low power eye-piece.

This remarkable condition seems, however, to be only temporary, and in fact often escapes observation. After a short time, which in about 100 observations was found to vary from a few seconds to a quarter of an hour, the apparently homogeneous blood-cell masses break up anew into individual corpuscles, which then continue isolated or in bead-like rows, but remain spheroidal, *i. e.*, do not regain their biconcal shape.

Those corpuscles which arrange themselves in rows present an appearance strikingly different from the ordinary rouleaux arrangement of normal blood-disks, an appearance which may better be designated as "beaded," because the corpuscles are here spheroidal and not disk-like.

Liquids in general variously modify the shape of the red blood-disks, but no liquid or reagent tried in control experiments produces the effects described above.

Watery solutions of dried venom did not exhibit the immediate influence upon the red blood corpuscles as well as the fresh venom, although the corpuscles very promptly became spheroidal as they do from most watery liquids; but did not lose the coloring matter as when exposed to pure water without venom.

The blood of birds, upon being mixed with venom, does not show the above described changes in as striking a manner as mammalian blood. The nuclei of the oval corpuscles of the pigeon appear, however, to undergo a rapid necrotic change which finally gives rise to a granular albuminoid material to be seen floating in large quantities between the corpuscles.

A number of experiments were made to study the changes in the corpuscles of the living animal. Fresh venom or solutions of dry Crotalus venom were injected hypodermatically, and then the blood taken at intervals from the local lesion as well as from the general circulating fluid and examined under the microscope.

The blood taken from local lesions presented quite often alterations in the corpuscles similar to those observed in a direct mixture of blood and venom under the microscope, as described above. It was not possible, however, to trace all the modifications of the red blood corpuscles in specimens of the circulating blood; only one change being constant, viz., the spheroidal transformation of the blood disks. The red blood corpuscle retained the acquired spherical shape after the death of the animal.[1]

All the experiments made in order to study the ultimate changes in the blood-corpuscles gave nearly similar results varying slightly in degree with the quantity of the venom and the animal employed. The record of one observation will suffice for all.

Experiment.—Young cat. Injected at 2 P.M. 3 m. m. fresh Crotalus venom in left thigh. Hair of part being previously clipped away.

2 minutes later. Animal well. Blood microscopically examined. Local lesion; blood-disks assuming spherical-shape. Blood from auricular artery showed no changes in the blood-disks.

5 minutes later. Animal well. Local lesion; blood-disks all spherical showing also gelatinoid behavior and ductility on pressure. Auricular artery, blood-disks normal.

8 minutes later. Animal restless. Local lesion shows only spherical shape of red blood-disks. Auricular artery also shows partial change of red blood-disks to spherical shape.

12 minutes later. Animal ill. Blood of local lesion as well as blood taken from jugular vein shows same changes as when last examined.

---

[1] Errors in distinguishing the spheroidal shape of the red blood-corpuscles from the round shape which the normal disk-like corpuscles exhibit when viewed in a certain position are easily eliminated when the blood is brought into current by gentle pressure upon the cover-glass, or by inclining the stage of the microscope. The disks assuming spheroidal shape are decidedly reduced in diameter and appear smaller.

PATHOLOGY.

20 minutes later. Animal quite ill. Red blood-disks all spheroidal when examined in the local lesion and in several other parts of the body.

25 minutes later. Animal dead.

30 minutes later. Blood examined from heart. All the red blood-disks spherical.

24 hours later. The dead animal being kept in a cool place. Red blood-disks all spherical and many disintegrated.

The blood, in sections of tissues from animals poisoned with venom, also presents decided alterations. The corpuscles in tissues hardened with preserving fluid are seldom seen intact. When, however, the parts in question are placed in preservative fluid immediately after the death of the animal the spheroidal (altered) red blood corpuscles may be distinguished; and still more likely are they to be intact if the animal was killed before the venom had asserted its fatal effect.

The corpuscles as a rule appear disintegrated in animals dead from slow Crotalus venom poisoning, and present themselves as a granular débris of a yellowish or dark brown color. The tissue elements of the part into which the venom had been directly injected are as a rule profusely saturated with the coloring matter of the blood. The microscope further reveals blood crystals and numerous bacteria between and within the tissue elements. This indicates the profound alteration which takes place in the blood in venom poisoning, and accounts for the black appearance and the rapid putrefactive changes which are seen in the local lesion.

"Quite recently Lacerda, in lectures[1] on snake poisons, speaks of alterations of the blood, which differ much from those observed by us. He does not state the serpent venom employed. It was presumably from the *Bothrops urutu*, *Lacer*. In slow poisoning, he says, the blood globules become indented like a toothed wheel. Some are elongated, deformed, or broken up; others present shining points and then break up into minute fragments. Some undergo a change of tint to chesnut brown, others become entirely discolored.

"The consequences of mixing pure blood and pure venom, he says, are these: The red blood-globules unite in mass, adhere one to the other and begin thereon to lose their normal forms. In a few minutes the dissolution is complete. There remains only an amorphous protoplasmic matter, semi-liquid, diffluent, of a uniform yellow color, with well-marked red striations. After some minutes the hematine or coloring matter quite disintegrated is seen under the form of granular substance of a deep vermillion red. Whilst the globules thus break up bubbles of gas rise here and there."

We have quoted this account nearly in full to point out that it describes a sequence of appearances very unlike those which we have delineated.

Effects of the Venom upon certain Tissues.—Direct observations were further made as to the effects of venoms upon the various solid tissues, such as the bloodvessel walls; muscular tissue, unstriated and striated; nervous tissue (brain and medulla

---

[1] Leçons sure le Venin des Serpentes du Brésil, etc. J. B. De Lacerda, pp. 87–88. Rio de Janeiro, 1884.

oblongata); lungs, liver, skin, mucous membranes, the cornea, spermatozoa and ciliated epithelium, and most extensively upon the mesentery and other serous membranes.

If fresh venom be injected into any organ or applied to any internal part of the body, one of the chief effects is, as Dr. Weir Mitchell showed twenty-two years ago, the production of minute hemorrhages.

The studies reported here thoroughly confirm Dr. Mitchell's observations. Above all, it was further evident that, as a general rule, *it was everywhere the parenchymatous elements of the organ or parts that underwent necrotic changes (which will be described below), while the interstitial elements of organs or tissues acted upon by the venom, remained usually unaffected, or were merely infiltrated with blood, or with the disintegrated products of blood.*

Effects of Venom upon Bloodvessel Walls.—If fresh venom be applied to a vascular tissue and watched under the microscope, no effect upon any of the larger bloodvessels is perceptible. It appears that the venom even after prolonged contact has no visible influence upon the smooth muscular tissue constituting the middle coat of arteries and veins. The adventitia of such vessels also offer considerable resistance, although the vasa vasorum become extremely congested. Even small arterioles and venules are unaffected.

The capillary bloodvessels, however, having a mere endothelial wall surrounded by a delicate adventitia show a decided change upon the application of venom. The endothelium constituting the capillary wall becomes cloudy and looks as if roughened and displaced, and though no actual rupture of the vessel is demonstrable to the eye a diapedesis of blood-corpuscles and a leakage of serum occurs, and this process is sometimes amazingly rapid. As will be described later with more details, the following are the essential points in the action of the venom upon its direct application to a vascular membrane viewed under the microscope. The blood current appears at first to be accelerated, and the color of the blood becomes darker. Then in a few moments while the circulation still continues in the veins and arteries, in many of the capillaries stagnation occurs. From these latter vessels, and apparently only from them, the blood oozes, first forming pin-point ecchymoses which gradually increase, and which by fusion give rise at last to a general hemorrhagic infiltration of the neighboring tissues.

Changes in the Striped Muscular Tissues.—Venom directly applied to living striped muscular tissue produces changes which become apparent immediately after the death of the animal. The ultimate muscular fibrillæ readily break up into their sarcous elements, these becoming easily separable in transverse layers, the so-called Bowman's disks. A granular change[1] of the elements, not however uniform in character and distribution, is quite conspicuous. (Fig. 3.) It should be noted that all these changes occur without the addition of any other reagent than venom and become conspicuous when the poisoned tissue is teased out in water.

The alterations just referred to are most manifest in muscular fibres near or around which the capillaries are affected by the venom, localities apt to be marked

---

[1] Figured by Dr. Mitchell, in his first essay.

by the presence of extravasated blood. The changes described are not uniform, even in an individually affected fibre, but are bounded by small abrupt layers of

Fig. 3.

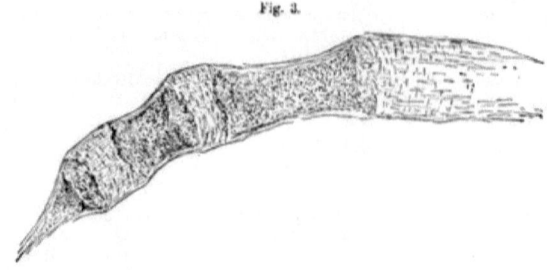

normal sarcous elements, the whole being surrounded by an unaffected sarcolemma. The latter, which is beautifully demonstrated on such occasions, shows constrictions in those places where the sarcous elements are disintegrated. (See Figs. 3 and 4.)

Fig. 4.

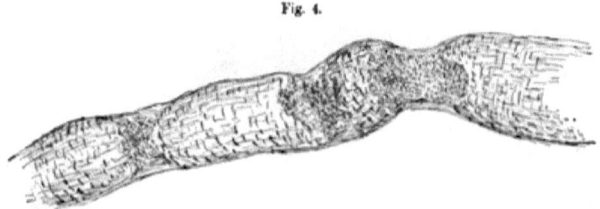

All muscular fibres of the part where the poison was injected are more or less granular, and are often stained by hematin. The granular material between the fibres has very much the appearance of micrococci, but by appropriate tests only a small proportion of the granules can be identified as bacteria. The remaining granular matter can be identified as particles of necrosed sarcous elements, disintegrated blood, and granular substances which have been described as constituents of the fresh venom and introduced from without. These granular muscle changes occur only in or near the wound, and not also in remote muscles. They demand for their production a certain length of time, and are most decided in cases of long survival.

Changes in the Lungs.—As has been seen from the table of experiments, the injection of venom into the lungs was followed by nearly instantaneous death. The local lesion was an hemorrhagic infarction throughout the whole of the parenchyma, filling also all of the air vesicles. There were extensive sub-pleural ecchymoses, both parietal and visceral, as well as sub-pericardial ecchymoses. Under the microscope with high amplication sections of the lung tissue showed a peculiar clogging, fusion, and ductility of the red blood-corpuscles not unlike that which follows the application of fresh venom to the blood.

This appearance is, however, not uniform, as in many places the corpuscles are merely spheroidal, or have undergone a granular disintegration. The bloodvessels are all highly congested, the air vesicles seem to be distended by extravasated blood. The micrococci introduced with the venom appear to have rapidly multiplied, numerous masses being seen in the air vesicles and in the more necrosed parts of the lung tissue. In general the tissue is deeply stained by the coloring matter of the blood.

Brain and Medulla Oblongata.—If a minute quantity of venom was successfully injected directly into the cranial cavity of a pigeon, the animal fell immediately and expired in a few minutes. The pia mater was preëminently the seat of hemorrhages, and blood was also seen to fill the peri-vascular spaces of many of the cerebral vessels. The cerebral tissue, and particularly the nerve elements, showed a granular change analogous to that observed in muscular tissue, and similar appearances were noted from the effects of the venom upon the medulla oblongata and spinal cord. Where animals were poisoned by introducing the venom subcutaneously into some other part of the body, minute capillary hemorrhages were also observed in the membranes of the brain, and in two instances ecchymoses were noted in the substance of the medulla oblongata, although as a rule only intense congestion of its vessels was seen. The bloodvessels were so much distended with blood as to be double or even triple their normal calibre, fully obliterating the perivascular spaces and unquestionably exerting much pressure upon the surrounding nerve elements.

Effects of the Venom when applied to Uninjured Mucous Membranes, and upon the Cornea.

The following experiments were made:—

Experiment.—Adult albino rabbit, etherized. A drop of an aqueous solution of the dried venom *Crotalus adamanteus* was dropped on the cornea and conjunctiva of the left eye. In a few minutes the conjunctiva became ecchymosed and œdematous to such an extent as to close the eyelids. Animal died in five hours. After death the conjunctiva and eyelids were seen to be soaked with extravasated blood, while the cornea remained perfectly transparent and colorless, showing no trace of inflammatory change when removed and examined under the microscope.

Post-mortem examination showed extensive sub-pleural, sub-peritoneal, and slight sub-arachnoid ecchymoses.

Experiment.—Young kitten, etherized. A drop of fresh venom was placed on the cornea. Results, similar to those of foregoing experiments, the cornea remaining transparent, but exhibiting a certain roughness upon the surface, which under the microscope proved to be due to slight desquamation of the epithelium.

Experiment.—Young kitten, etherized. Abdominal cavity and stomach opened and fresh venom applied to surface of mucous membrane. Specimen watched for half an hour failed to reveal any decided visible changes, beyond a slight corrugation and congestion. No ecchymoses.

The Effect of Crotalus Venom upon Ciliary Motion.—Fresh venom applied to ciliated epithelium taken from the edge of the tunic of a fresh oyster seemed to exert no effect upon ciliary motion. The specimen was watched and compared with the control experiment side by side. The ciliæ still kept up their movement at the end of three hours in both specimens.

Fresh venom was applied to ciliated epithelium taken from the pharynx of a live frog. The specimen was carefully observed and compared with similar preparations in which venom was not used. In the latter the ciliary motion, as a rule, kept up longer. Yet after one hour specimens treated with venom continued to exhibit motion though less vigorous than in the control specimens.

The Effect of Venom upon Spermatozoa.—Fresh venom applied to spermatozoa taken from a live rabbit seemed to exert a decided influence. Specimens treated with the venom were examined side by side with control specimens, and while in the presence of venom the spermatozoa ceased to exhibit their peculiar movements in from one-quarter to three-quarters of an hour, unpoisoned spermatic particles continued to move for many hours. The venom did not appear to produce any changes in the substance or the bodies of the individual spermatozoa.[1]

The Mechanism of the Hemorrhages as Observed in Venom Poisoning.

In order to study the mechanism of the hemorrhages Dr. Mitchell's original observations were repeated as follows:—

The animals used were cats, rabbits, pigeons, white rats, and frogs. The frogs do not give satisfactory results as they withstand the effects very strenuously and if peritoneal hemorrhages occur at all they are very scanty. The most satisfactory observations were obtained when cats were employed, as these animals lived longest after the application of the venom, the latter also acting more slowly, thus permitting satisfactory study of the effects under the microscope.

Anæsthetics were always used. Ether was found to give the best results. Chloral appeared to retard the effects of the venom. While in an etherized animal peritoneal hemorrhages appeared at once upon the application of the venom, in a chloralized animal they occurred much later, and sometimes failed to appear.

A few drops, three to six, of a saturated solution of chloral hydrate were usually sufficient to anæsthetize a small kitten or rabbit, two drops for a white rat, one drop for a mouse. It was administered hypodermatically. In administering ether the animal was placed under a bell-glass, with a sponge kept saturated with the agent until the animal was rendered powerless.

In experiments upon the mesentery to be examined under the microscope, the

---

[1] The observations were made under an amplification of one thousand diameters.

animal was placed on its right side upon a thin oblong wooden board. On one side of the board near the middle was cut a triangular opening, each side being about one inch in length. An incision was then made in the median line through the abdominal integuments, sufficiently large to allow a loop of the intestine to be extracted. Care was taken that pressure should not interfere with the circulation. The loop being drawn out, it was stretched over the hole in the board above described, and kept in position by means of pins.

The venom was applied to the uninjured surface of the mesentery. A saturated aqueous solution of the dried venom was most commonly used. The moist chamber was not required, as the experiments were of short duration. The warm stage seemed only to hasten the process and otherwise was observed to have no special influence, being rather disadvantageous.

Experiment 1.—A young kitten was secured by means of ether as above described, and placed upon the microscopic stage. A few drops of an aqueous solution of the venom were allowed to flow over the mesentery. The part being carefully watched with the naked eye, it was noticed that after one minute tiny hemorrhagic points made their appearance here and there, all over that part of the mesentery which was under the direct influence of the venom. These hemorrhages increased rapidly in size, and in a few minutes the whole surface became the seat of one diffused hemorrhagic infiltration. (Plate III., Figs. 3, 4, 5, 6, 7.)

Experiment 2.—Young white rat. Ether. Aqueous solution of venom applied as before to the mesentery. The loop of the mesentery acted upon was quickly cut out by means of scissors after the lapse of one minute and subjected to drying. A beautiful preparation was thus obtained, in which the minute hemorrhages were permanently fixed by drying, preserving their natural appearance.[1]

Experiment 3.—Young kitten. Chloral. Mesentery spread upon microscopical stage. An aqueous solution of dried venom applied in the same manner as above. In this case the hemorrhages did not appear so promptly and were not so rapid in their development. Nearly five minutes elapsed before they began to form.

Experiment 4.—Young white rat. Chloral. Venom applied as in preceding experiments; there seemed to be delay in the appearances and development of the hemorrhages.

Further enumeration of this class of observation is unnecessary, as more than forty experiments exhibited the characteristic hemorrhages, except in the case of frogs. Five of the latter were used.

It was noted that chloral always retarded the production of hemorrhages, at least they did not appear as rapidly as when ether was used.

Microscopical Details.—It being necessary to study the exact location of the hemorrhages and the mode of the escape of blood, the following modifications of methods were adopted in repetition of the older experiments of Dr. Mitchell.

---

[1] After much experimentation this method of preparing permanent specimens was found to be the only available one. Specimens of mesentery mounted in any kind of liquid very soon lose their proper appearance, as the hemorrhagic specks in the membrane gradually vanish, or get blurred from the effects of the preserving fluid.

Experiment 5.—Cat. Ether. Mesentery exposed and placed on the stage of the microscope. The aqueous solution of venom was applied, and the experiment watched under a magnifying power of 60 diameters. In thirty seconds minute hemorrhagic points were noticed as in all the previous experiments first along the sides of the smallest capillaries. It was also observed that the hemorrhages occurred first in those small capillaries which were in the neighborhood of larger vessels. In vascular plexuses which started from the greater arteries the hemorrhages appeared much sooner than in those which took their departure from smaller arterioles. In each case, however, it was only the capillaries from which the hemorrhages proceeded, the arteries and veins remaining intact. The hemorrhages being seen to proceed from capillaries in the vicinity and along the route of larger vessels, one may erroneously get the impression that it is the latter from which the bleeding arises. No actual breech of continuity in the capillaries was observed, and it appeared as though the blood filtered through the walls of these minute channels.

Experiment 6.—Kitten. Ether. Mesentery exposed in the usual manner. The mesenteric vessels, both the main artery and the veins, were ligated near the root of the mesenteric attachments. A salt solution stained by aniline blue was injected into the vein. The venom was applied, and the field closely watched under the microscope. No extravasation of the injected solution or of blood could be observed.

Experiment 7.—Kitten. Ether. Vessels ligated and salt solution with aniline injected as in previous experiment. Applied aqueous solution of the dry venom. No extravasation of the colored liquid or blood observed.

Experiment 8.—Kitten. Animal secured as before, but no salt solution injected. Mesenteric veins and the artery ligated near root of mesentery. Solution of dried venom applied. Hemorrhage as usual, but slow and scanty.

Experiment 9.—Kitten. Ether. Animal fixed as in last experiment upon microscopical stage. Fresh venom applied and watched for one-half hour. Hemorrhages were seen to develop more slowly.

Experiment 10.—Kitten. Ether. Mesenteric vein and artery ligated as in preceding experiments. Fresh venom applied as before, and a marked delay in development of hemorrhages again observed.

Experiment 11.—Kitten. Ether. Mesenteric vessels tied not only at the root of the mesentery, but also peripherally at the convex portion of the loop, thus almost entirely cutting off the circulation. Fresh venom applied. Hemorrhages were scarcely appreciable with the naked eye.

Experiment 12.—Kitten. Ether. Mesenteric vessels tied at both root and periphery of mesentery. Venom applied immediately. Hemorrhage hardly perceptible.

The above experiments were subsequently repeated, especially those in reference to the effects of the venom upon bloodvessels when blood had been substituted by a 0.75 per cent. saline solution. These experiments, however, gave the same results as those just described, and hence it is unnecessary to occupy space in multiplying similar records. Yet some studies in the same direction have been left undone. It might be desirable to elaborate the methods of experimentation, *e. g.*, by application of artificial blood pressure, etc. In all the experiments above quoted, except in experiments 6 and 7 (where the blood was substituted by another liquid), the peculiar extravasations of blood followed the application of the venom.

When the mesenteric vessels were tied as in experiments 8, 9, and 10, there was a delay in the appearance of the hemorrhages. When the vessels were tied in two places, as in experiments 11 and 12, so as to cut off the circulation in a great measure, the hemorrhages appeared hardly appreciable to the naked eye. And as we have seen above, there was no extravasation at all when the blood was substituted by an artificial liquid, as in experiments 6 and 7.

Therefore, the *hemorrhages become less marked in proportion to the interference with the circulation of the blood in the part.*

CHAPTER XI.

GENERAL CONSIDERATIONS.

It seems desirable at the close of a research such as we here record to offer a few brief and general considerations in connection with some of the methods and plans pursued in parts of the work, to group some of the conclusions, and to bring together deductions which are necessarily scattered. A summary is also desirable that we may set forth succinctly the essential actions of venom so as to make clear the important differences in the toxic influences of globulins and peptones, to facilitate the application of what we have learned to the treatment of snake bite, and to indicate new lines of research in the most promising directions.

Our discovery of the existence of two distinct classes of poisons in venoms, that both are doubtlessly represented in all venoms, only differing in relative proportions and slightly in chemical and physiological properties, that they possess activities akin but yet readily distinguished, and that they are proteids and closely related to principles normally existing in mammalian blood, seems to us as of grave importance. Our methods, however, for the separation of the poisonous substances in venoms are open to improvement, because the processes are slow, and since possibly one of the poisons at least is injured. It does not seem from the results of our physiological studies with these poisons that any of them excepting the *copper-venom-globulin* have suffered, but that this has been affected seems probable from its altered solubility, its comparatively low toxic power, and its physiological peculiarities compared with the other globulins. Doubtless the ordinary methods for the separation of the globulins from other proteids in solution could be used to advantage, but how far successful they may prove in isolating the globulins from each other can only be determined by extended and careful investigation.

The plan we adopted in studying venoms and their active principles on the arterial pressure, pulse, and respiration is probably open to much criticism, but any other course seemed unavoidable. Instead of studying all venoms together as though they were absolutely identical compounds, although from different sources, and each of the active elements, as for instance the peptones, together as identical, it would doubtless have been preferable to have made a detailed investigation of each venom, and of each of the active principles of that specimen. But this course could not have been pursued satisfactorily because of the meagre supply of poison. It was then simply a question as to whether we would take a very limited number of experiments with each venom and each of its active principles, and base conclusions thereon, or study the actions of all pure venoms together, of all the water-venom-globulins together, etc., and then form our conclusions. The latter course seemed preferable: first, because of the similarity in the actions of all pure venoms and of the ready interpretation of any differences, and of the resemblance in the actions of members of each of the classes of poisons; second, because in some of the actions such diverse factors are at work as to give apparently contradictory results, so that conclusions founded upon a very limited number of experiments would likely be more misleading than in the plan we adopted.

We summarize the following important points, deduced chiefly from our studies of Crotalus venom, to which are added a few comments:—

1. Venoms bear in some respects a strong resemblance to the saliva of other vertebrates.

2. The active principles of venom are contained in its liquid parts only. The solid constituents, such as we observed suspended in the poison, consist of epithelium cells, some minute rod-like animal organisms and micrococci, etc., which, when separated from the liquid fresh venom by means of filtration and well washed by water are harmless. Micrococci are constantly present in fresh venom, but have nothing to do with its virulence.

3. Venoms may be dried and preserved indefinitely in this condition with but very slight impairment of their toxicity. In solution in glycerine they will also probably keep for any length of time.

4. There probably exist in all venoms representatives of two classes of proteids, *globulins* and *peptones*, which constitute their toxic elements; the former may be represented by one or more distinct principles.

5. When venom is taken into the stomach in the intervals of digestion, enough of the poison may be absorbed to produce death, especially in the case of those venoms which contain a larger proportion of the more dialysable peptone; but during active digestion the venom undergoes alteration and is rendered harmless.

6. Potassic permanganate, ferric chloride in the form of the liquor or tincture, and tincture of iodine seem to be the most active and promising of the generally available local antidotes.

7. Venom exerts a powerful local effect upon the living tissues, and induces more rapid necrotic changes than any known organic substance. It causes œdema, swelling, attended with darkening of the parts by infiltration of incoagulable blood, breaking down of the tissues, putrefaction, and sloughing.

8. It renders the blood incoagulable.

9. When brought in contact with a vascular tissue of a warm-blooded animal it produces such a change in the capillary bloodvessels that their walls are unable to resist the normal blood pressure, thus allowing the blood-corpuscles to escape into the tissues. These lesions are, however, not analogous to those of inflammation, since in the latter process it is principally the white blood-corpuscles which emigrate from the vessels, and the blood is highly coagulable, while here the blood exudes *en masse* and coagulates with difficulty, if at all. Free access of air (probably of oxygen) appears to lessen the virulent effects. The mesentery exposed to air, and on which the venom is merely brushed, endures the venom longer and in much larger quantity than when the poison is injected into the unopened and

GENERAL CONSIDERATIONS.

uninjured peritoneal cavity, or when directly thrown into the blood. There may be here also a question of temperature and other conditions.

The following facts as elicited in these investigations seem to be sufficient to explain the mechanism of the hemorrhages: the blood pressure has been shown to play a most important part; a watery salt solution substituted for the blood does not extravasate, hence, blood seems to be necessary; there always occur molecular changes in the bloodvessel walls from the effect of venom. That blood pressure is an important factor has been established by the observation that the hemorrhages as a rule occur first in the capillaries which are immediately next to or nearest the large bloodvessels. The hemorrhages take place soonest where the force of the blood current is first felt and cannot be sufficiently resisted, and in no case do hemorrhages seem to originate from vessels with strong walls like the arterioles or veins. Cutting off the circulation of a part, as, for instance, by ligation of the vessels of the mesentery, destroys the blood pressure, and, as a consequence, the hemorrhages are so slight as scarcely to be seen by the naked eye though venom was freely applied. Finally, the colloid, softened, diffluent condition of the red corpuscles must inevitably facilitate extravasations. It is impossible to have seen numerous cases of venom poisoning without noting a variety of symptoms often abrupt or unexpected. These often are due, as Dr. Mitchell long since pointed out, to accidental hemorrhages into brain, kidney, and heart tissues. They explain much which might otherwise seem inscrutable, and serve sometimes to give a marked individuality of symptoms to cases which survive long.

10. Among the most remarkable effects of venom is that upon the red blood-corpuscles. These bodies undergo substantial modifications, i. e., they lose their bi-concave shape, become spherical and softened, and fuse together into irregular masses acting like soft elastic colloid material. This jelly-like condition of the corpuscles is no doubt doubly important: in connection with the extravasation of the blood, and in its probable interference with the normal respiratory functions of the blood-cells.

11. The direct action of venom upon the nervous system save as concerns the paralysis of the respiratory centres is of but little importance.

12. The alterations in the pulse-rate are dependent chiefly upon two antagonistic factors which are active at the same time, the one tending to increase the rate and the other to diminish it. The former is found in the increased activity of the accelerator centres and the other in a direct action on the heart. When we have the action on the accelerator centres removed by isolation of the heart from any centric influence we almost invariably find a diminution of the heart beats. Occasionally after this operation the pulsations are increased, but this alteration is attended, as in the case of the diminution of the pulse, by feeble heart beats, and accordingly is but a manifestation in another way of a depressed condition of the heart.

13. The variations in arterial pressure are due chiefly to three causes, depression of the vaso-motor centres, depression of the heart, and irritation and consequent constriction or blocking up of the capillaries. It seems not improbable that all of these are consentaneously active, and it therefore follows that such alterations are dependent upon the relative degree of power exerted by any one of these factors. Our

results indicate that the profound primary fall of arterial pressure is chiefly due to depression of the vaso-motor centres and is in part cardiac, that the subsequent recovery is capillary, while the final fall is cardiac. The initial fall does not continue, because the constriction of the capillaries is, for a time at least, capable of compensating the depressed action of the central organ of circulation.

14. The respirations are primarily increased and secondarily diminished. Here again we have two antagonistic factors at work together, one tending to increase and the other to diminish the rate. The former is an irritation of the peripheries of the vagi nerves and the latter a depression of the respiratory centres; whether we have an increase followed by a decrease or a decrease from the first will depend upon the relative intensity of the action of the venom on these two parts. When the action of the venom is sufficient to profoundly depress the centres the excitation of the peripheries may prove futile.

15. Death in venom-poisoning may occur through paralysis of the respiratory centres, paralysis of the heart, hemorrhages in the medulla, or possibly through the inability of the profoundly altered red corpuscles to perform their functions. There can be no question, however, that the respiratory centres are the parts of the system most vulnerable to venom, and that death is commonly due to their paralysis.

A general survey of the chief physiological actions of venoms leads us to believe that the most important effects are upon the respiratory and circulatory apparatuses, and that in the production of these results antagonistic factors are at work so that we sometimes have observations which seem directly contradictory. When it is remembered that there are two classes of poisons in venoms, that each class possesses certain distinguishing physical, chemical, and physiological differences, although closely related, it is easy to conceive of the cause of the existence of antagonistic actions and the necessarily varying results.

A comparative study of the actions of the globulins and peptones indicates that the *globulins* produce swelling and blackening of the parts by infiltration of incoagulable blood; they are the more potent in producing ecchymoses, in destroying the coagulability of the blood, in modifying the red-corpuscles, and in the production of molecular changes in the capillary walls; their action on the accelerator centres of the heart is more notable than that of the peptone, hence they are more active in causing the increased pulse-rate; they exert, too, a more marked action on the vaso-motor centres in producing the primary fall of pressure and are the greater depressants of the heart; they also act more powerfully upon the respiratory centres to paralyze them. The *peptones* are more active in the production of œdema, in the breaking down of the tissues, in the production of putrefaction and sloughing; they have little power to produce ecchymoses, to prevent coagulation or modify the capillary walls or the blood-corpuscles; they have less tendency to accelerate the pulse; they tend to increase the blood-pressure by irritating the capillaries, and are the principal factor in exciting the peripheries of the vagi nerves in the production of the increased respiration-rate.

A knowledge of these peculiarities in the actions of *globulins* and *peptones* coupled with the fact that the two classes exist in different proportions in the various species of venoms is of great importance in explaining the diverse pathological appearances in cases of poisoning in different kinds of snake-bite—and suggests immediately the cause of the frightful local changes which are seen after the bite of the Crotalidæ but scarcely at all in Cobra-poisoning. It must not, however, be supposed that the peptones or globulins for instance are absolutely identical physiologically in every venom, as they are probably modified physiologically as well as chemically, although we do not doubt that on the whole the type of action is carried throughout all species. Cobra venom does not produce the marked lesions of Crotalus-poisoning because it is so lacking in globulins; it is weak in the production of the local swelling and blackening of the parts, of the ecchymoses, of the altered corpuscles and of the non-coagulability of the blood, but the effects of Cobra venom are closely in accord with the actions peculiar to peptones. The peptone of Cobra seems to have a more decided power in producing convulsions than that of the rattlesnake.

The fact that the active principles of venom are proteids, and closely related chemically to elements normally existing in the blood, renders almost hopeless the search for a chemical antidote which can prove available after the poison has reached the circulation, since it is obvious that we cannot expect to discover any substance which when placed in the blood will destroy the deadly principles of venom without inducing a similar destruction of vital components in the circulating fluid. The outlook then for an antidote for venom which may be available after the absorption of the poison lies clearly in the direction of a physiological antagonist, or, in other words, of a substance which will oppose the actions of venom upon the most vulnerable parts of the system. The activities of venoms are, however, manifested in such diverse ways and so profoundly and rapidly that it does not seem probable that we shall ever discover an agent which will be capable at the same time of acting efficiently in counteracting all the terrible energies of these poisons.

It is now most desirable that our discovery of the complex chemical nature of venoms should be made the groundwork in India of a study of the poison of the Bungarus, the Daboia, and especially of the dangerous Hydrophideæ. So far all efforts on our part to obtain the venoms of Australian and South American snakes have failed, so that these and the dreaded Vipère Fér-de-lance and the large Elaps of Mexico remain so far really unstudied.

BIBLIOGRAPHY

Abstract of report of the Commission appointed at Calcutta to investigate the subject of snake poisoning. Rep. on San. Meas. in India, Lond., 1875, VII, 166–169.

ADAMS (F.). Case of a woman stung by an adder. Month. J. M. Sc., Edinb., 1841, I, 796–797.

AGAZZI (A.). Morte per morsicatura di vipera. Gazz. med. ital. lomb., Milano, 1851, 3. S., II, 211.

ALBA. Cas de morsure de vipère traitée par les injections d'ammoniaque et terminée par la mort. Ann. Soc. méd. de l'arrond. de Neufchateau, 1874, I, 13–15.

ALBERTONI. Sull' azione del veleno della vipera. Sperimentale, Firenze, 1879, XLIV, 142–153. Also: Ann. di chim. applic. a. med., Milano, 1879, LXIX, 210.

ALDRIDGE (E. A.). Snake bite, inflicted by the bamboo snake. [2 cases.] China. Imp. Customs. Med. Rep., Shanghai, 1882, No. 22, 9.

Ammoniacal injections in snake bite. Austral. M. Gaz., Melbourne, 1869, I, 260.

ANDERSON (E. A.). On the use of bromide of potassium in rattlesnake bites (*Crotalus horridus*). Am. J. M. Sc., Phila., 1872, Apr., 366–368.

ANSELMIER. Morsure de vipère chez un saltimbanque algérien en représentation à Paris. Rev. méd. franç. et étrang., Par., 1868, I, 229–235.

APPIAH (C.). A case of snake bite. Madras Month. J. M. Sc., 1873, VII, 334–336.

ARON (T.). Experimentelle Studien über Schlangengift. Centralbl. f. klin. Med., Leipz., 1882, III, 481–484. Also: Ztschr. f. klin. Med. Berl., 1883, VI, 332; 385.

Ash tea as a remedy for the bite of a rattlesnake. Nashville J. M. and S., 1858, XV, 471.

Azione (Intorno all') azione del veleno viperino sul sangue, ed à suoi contravveleni. Gazz. med. ital., prov. venete, Padova, 1859, II, 129–132.

BÄCK (A.). Ondezoek aangaande den beet der slangen, en deszelps meerdere of mindere gevaarlykheid. Geneesk. Verhandel. a. de k. Sweed. Acad. (Sandifort), Leiden, 1775, I, 222–230.

BADALONI (G.). Sul valore del permangnanato di potassa quale antidoto del veleno dei serpenti (ofidi). 8°. Bologna, 1882. *Repr. from:* Bull. d. sc. med. di Bologna, 1882, 6. S., IX.

BADALONI (G.). Sul valore del permanganato di potassa quale antidoto del veleno dei serpenti (ofidi); rapporto. Bull. d. sc. med. di Bologna, 1882, 6. S., IX, 5–19.

BADALONI (G.). The poison of the viper, and permanganate of potash. Communicated to the Med. Soc. of Lond. in an English translation by Dr. I. Owen. Lancet, Lond., 1883, I, 768–770.

BADDELEY (P.). A case of snake bite successfully treated by bloodletting and cold effusion. India J. M. and Phys. Sc., Calcutta, 1836, N. S., I, 289.

BAILLIE (A.). Bite of a rattlesnake. Med. Times, Lond., 1849, XX, 179.

BAJON. Observation sur une morsure de serpent gnérie par l'usage de l'alkali volatil. J. de méd., chir., pharm., etc., Par., 1770, XXXIII, 146–148.

BAKER (T. E.). Treatment of snake bites. India J. M. and Phys. Sc., Calcutta, 1836, N. S., I, 493.

BARDY (H.). De la morsure de vipère comme cause de mort. Bull. gén. de thérap., Par., 1875, LXXXVII, 502.

(159)

BARKER (E.). [Snake bite; autopsy.] Austral. M. J., Melbourne, 1860, V, 146-148.

BARNETT (J. M.). Case of snake bite. Tr. M. and Phys. Soc. Bombay (1857-8), 1859, N. S., IV, App., xxxvii.

BARON. Des serpents à sonnettes, de leur morsure, des effets qu'elle produit et des moyens d'y remédier. Clinique d. hôp. et de la ville, Par., 1827, I, No. 57, pp. 2-4.

BARTON. A case of a full-grown rabbit being bitten, two or three times, by a small rattlesnake; gradual recovery. Phila. M. and Phys. J., 1805, I, 167.

BARTON (B. S.). An account of the most effectual means of preventing the deleterious consequences of the *Crotalus horridus*, or rattlesnake. 4°. Philadelphia, 1792. *Repr. from:* Tr. Am. Phil. Soc., III.

BEARDMORE (F. J.). Report of a case of snake bite and recovery. Lancet, Lond., 1849, II, 288.

BEATTIE (W. F. C.). Recovery from bite of rattlesnake. N. York M. J., 1873, XVIII, 619.

à BECKETT (W.). Injection of ammonia in snake poisoning. Austral. M. J., Melbourne, 1868, XIII, 390-392.

BEDDOME (R. H.). Notes upon the land and fresh-water snakes of the Madras Presidency. Madras Q. J. M. Sc., 1862, V, 1-31; 1863, VI, 41-48; 1866, IX, 207; 1867, XI, 14-16.

BEHNCKE (G.). Forgiftning ved hugormebid. Hosp.-Tid., Kjbenh., 1877, 2. R., IV, 361-365.

BEHRENS (R. A.). Relation von einem Schlangen-Biss. Samml. v. Nat. u. Med.- [etc.] Geschichten, Leipz. u. Budissin. 1724, XXII, 453.

BENNETT (E.). Bite of an adder. Lancet, Lond., 1873, II, 102.

BERNCASTLE. Australian snake bites; their treatment and cure. 12°. Melbourne, 1868.

BERNCASTLE. On the distinction between the harmless and venomous snakes of Australia. Australian M. Gaz., Melbourne, 1870, II, 21.

BERNCASTLE. The treatment of snake bites. Austral. M. Gaz., Melbourne, 1869, I, 31-35.

BHUTTACHARJEA (S. C.). A case of snake bite; cure. Indian M. Gaz., Calcutta, 1873, VIII, 263.

BIANCURTTI (V.) Avvelenamenti iposteniei curati sulla scorta della dottrina medica italiana. Bull. d. sc., Bologna, 1859, 4. s., XI, 210-218.

BINZ. Ueber Indisches Schlangengift. Sitzingsb. d. nied.-rhein. Gesellsch. f. Nat.- u. Heilk. zu Bonn, 1882, 4, 169-171.

BISHOP (E.). On the poison of the adder. Lancet, Lond., 1862, I, 550.

Bite of a serpent. Med. and Surg. Reporter, Phila., 1868, XIX, 259.

Bite of a viper (*Coluber berus*) in the right hand. Lancet, Lond., 1829, II, 507.

BLACKBURN (J. C. C.). Bite of a rattlesnake; cure. North. Lancet, Plattsburg, N. Y., 1853, VII, 128.

BLANCHE. Spacèle du membre inférieur droit, suite d'un morsure de vipère à cornes. Rec. de mem. de méd. . . . mil., Par., 1864, 3. s., XII, 396-400.

BLANCHET. Morsure de vipère; guérison prompte par la succion suivie de cantérisation. Rec. d. trav. Soc. méd. d'Indre-et-Loire, Tours, 1852, 52-54.

BLAND (H.). Report of a case of snake bite with observations on the treatment in such instances. Lancet, Lond., 1848, I, 71.

BLAND (W.). Bites of the venomous snakes of Australia. Austral. M. J., Melbourne, 1861, VI, 1-6.

BLOT (JEAN-CHARLES). Dissertation sur la morsure de la vipère fer-de-lance. Thèse. VI, 7-35, pp. 4°. Paris, 1823, No. 106.

BOEHNER (J. B.) and BOEHNER (G. R.). De psyllorum, marsorum et ophiogenum, adversus serpentes eorumque ictus virtute. 4°. Lipsiæ, 1745.

BOGUE (R. G.). Rattlesnake bite; recovery. Chicago M. J., 1866, XXIII, 399-401.

BOILLE. Morsures de vipère, cas de mort; injection d'ammoniaque dans les veines; guérison. J. de méd. et chir. prat., Par., 1874, XLV, 450-452.

DE BOISMARMIN. Morsure de vipère; mort. Traitement à suivre. J. de med. et chir. prat., Par., 1875, XLVI, 350-354.

BIBLIOGRAPHY. 161

DO BOMFIRA (A. M.). Algnns apontamentos acerca das mordeduras das serpentes, e das picadas dos insectos venenosos. Gaz. med. da Bahia, 1869, III, 149; 184.

BONHOMME. Morsure de vipère rapidement mortelle. Rev. méd. de Toulouse, 1867, I, 182–184.

BOOTH. Account of a South American remedy for the bite of poisonous reptiles and rabid animals denominated algalia, yerba-del-sapo, or contra culebra. N. Eng. J. M. and S., Bost., 1814, III, 322.

BORIES (E.). Rattlesnake poisoning treated by potassium permanganate. Polyclinic, Phila., 1883, I, 57.

BÖTTCHER (C.). Von Vipera berus gebissener Hund. Beitr. z. prakt. Heilk., Leipz., 1834, I, 172.

BOULLET (L. J.). Étude sur la morsure de vipère. Thèse. 4°. Paris (1867).

BOYD (J.). Case of adder bite. Edinb. M. J., 1861, VII, 324–326.

BOZEMAN (N.). A case of amaurosis, resulting from a snake bite. N. Orl. M. and S. J., 1849–50, VI, 739.

BRAUN (D. J. H.). Merkwürdiger Fall eines schnell tödtlichen Vipernbisses. Mag. f. d. ges. Heilk., Berl., 1831, XXXIV, 361–381.

BRETON (P.). Case of fatal snake bite. Tr. M. and Phys. Soc. Calcutta, 1825, I, 55–58.

BRETON (P.). Account of experiments with cobra, bora, and bungara—as to comparative virulence of their poison. Tr. M. and Phys. Soc. Calcutta, 1826, II, 170–180.

BRIGHTON (J. G.). Case of death from a bite of a snake, at sea. India J. M. and Phys. Sc., Calcutta, 1837, N. S., II, 766, 1 pl.

BRIGHTON (J. G.). Bite of a rattlesnake. Lancet, Lond., 1839, II, 929.

BROWNE (C. W.) and SMYTH (S.). Case of snake bite and recovery. Lancet, Lond., 1883, I, 716.

BRUNTON (T. L.) and FAYRER (J.). On the nature and physiological action of the poison of Naja tripudians and other Indian venomous snakes. Part I. Proc. Roy. Soc., Lond., 1872–3; XXI, 358–374; 1873–4, XXII, 68–133.

BRUNTON (T. L.) and FAYRER (J.). On the nature and physiological action of the poison of Naja tripudians and other Indian venomous snakes. Rep. on . . . san. improve. in India, 1872–3, Lond., 1873, 275–283.

BRUNTON (T. L.) and FAYRER (J.). On the nature and physiological action of the poison of Naja tripudians and other Indian venomous snakes. [Part II.] Rep. on . . . san. improve. in India, 1873–4, Lond., 1874, 332–362.

BRUNTON (T. L.) and FAYRER (J.). On the nature and physiological action of the Crotalus poison as compared with that of Naja tripudians and other Indian venomous snakes; also, investigations into the nature of the influence of Naja and Crotalus poison on ciliary and amœboid action, and on vallisneria, and on the influence of inspiration of pure oxygen on poisoned animals. Rep. on san. meas. in India, 1873–4; Lond., 1875, VII, 266–274.

BRUNTON (T. L.) and FAYRER (J.). On the nature and physiological action of the Crotalus poison as compared with that of Naja tripudians and other Indian venomous snakes; also, investigations into the nature of the influence of Naja and Crotalus poison on ciliary and amœboid action, and on vallisneria, and on the influence of inspiration of pure oxygen on poisoned animals. Proc. Roy. Soc. Lond., 1875, No. 159, 261–278.

BRUNTON (T. L.) and FAYRER (J.). Note on the effect of various substances in destroying the activity of Cobra poison. Proc. Roy. Soc., Lond., 1878, XXVII, 465–474.

BUGNION. Relation d'une morsure de vipère, suivie de guérison. Bull Soc. méd. de la Suisse Rom., Lausanne, 1879, XIII, 335–339.

BULL (G. H.). Recovery after the bite of a cobra. Indian M. Gaz., Calcutta, 1880, XV, 271.

BULLEN (F. D.). Snake bite. Tr. Cork M. and S. Soc., 1866–7, 85. Also: Dublin Q. J. M. Sc., 1867, XLIV, 492.

BURDER. [Case of death from bite of cobra.] Med. Times, Lond., 1852, N. S., V, 441.

BUTAZZI. Poisoning from a viper bite; cure with sulphate of quinine. Lond. M. Gaz., 1837, XIX, 591.

21   June, 1880.

BUTTER (DONALD). On the treatment of persons bitten by venomous snakes. Tr. M. and Phys. Soc. Calcutta, 1826, II, 220-233.

BUTTER (D.). Snake bite curable and hydrophobia preventible. 8°. London, 1873.

CANTOR (T. E.). Sketches of two undescribed venomous serpents, with fangs behind the maxillar teeth (à crochets postérieurs) [Cerberus Grantii and Potamophis]. Tr. M. and Phys. Soc. Calcutta, 1836, VIII, 137-142.

CARPENTER (G. H.). Iodine an antidote for snake bite. Med. News, Phila., 1883, XLII, 441.

Case of bite from a viper; severe symptoms; recovery. Lancet, Lond., 1852, I, 612.

Case (A) of snake bite. Month. J. M. Sc., Edinb., 1852, XV, 553.

Cases of snake bites. Tr. M. and Phys. Soc. Bombay, 1853, N. S., I, 314-316.

CASEY (C. G.). Case of snake bite. Austral. M. J., Melbourne, 1873, XVIII, 338.

CHALMERS (W.). Case of bite of a young woman by a large cobra de Capello. Glasgow M. J., 1853-4, I, 102-104.

CHAPPLE. Case of snake bite; death in an hour and a quarter. Lancet, Lond., 1873, II, 41.

CHARRIEZ (A.). De la piqûre du serpent de la Martinique. Thèse. 4°. Paris, 1875, No. 116.

CHATIN (H.). Des morsures de vipères et des indications curatives que présente le traitement de ces plaies envenimées. Rev. méd. franç. et etrang., Par., 1855, I, 129-136.

CHISINI (G.). Sul veneficio viperino. Gazz. med. ital., prov. venete, Padova, 1866, IX, 137-139.

CLARKE (H.). Case of snake bite. Tr. M. and Phys. Soc. Calcutta, 1829, IV, 442.

CLEGHORN (J.). Bite from a snake, said to have been poisonous; liquor ammoniæ treatment; recovery. Indian Med. Gaz., Calcutta, 1870, V, 131.

CLONA (J. N. D.). A case of snake bite. Eclect. M. J., Atlanta, 1879-80, I, 287.

CLOQUET (H.). Serpens venimeux. Dict. d. sc. méd. Par., 1821, LI, 175-189.

CLUTTERBUCK (J. B.). On two cases of snake bite successfully treated. Austral. M. J., Melbourne, 1864, IX, 195.

COCKLE (JOHN). An essay on the poison of the cobra di Capello. 32 pp. 8°. London, S. Highley & Son, 1852.

COCKS (T.). Case of viper bite, successfully treated. Lancet, Lond., 1827, XII, 201.

COLEMAN (R. T.). Case of poisoning by a snake bite. Virginia Clin. Rec., Richmond, 1872-3, II, 137-139.

COMFORT (A. I.). Case of snake bite. Phila. M. Times, 1878, IX, 77-79. *Also:* Hahnemann Month., Phila., 1879, N. S., I, 147-150.

COOKSON (H.). Case of bite of cobra; recovery. Indian M. Gaz., Calcutta, 1873, VIII, 321.

CORNISH (W. R.). Cobra bite; immediate treatment, recovery. Indian M. Gaz., Calcutta, 1880, XV, 271.

CORRE (A.). Analogie des symptômes et des lésions chez les individus mordus par des serpents venimeux et chez les individus empoisonnés par certains poisons. Arch. de phys. norm. et path., Par., 1871-2, IV, 405-408.

COSTELLO (C. P.). A successful case of venomous snake bite. Indian M. Gaz., Calcutta, 1869, IV, 92.

COUES (E.) and YARROW (H. C.). Notes on the herpetology of Dakota and Montana. 8°. Washington, 1878. *From:* Bull. U. S. Geol. and Geog. Survey, Vol. IV, No. 1.

COUTY (L.). O permanganato de potassa contra a mordedura de cobras. [With a reply by M. Lacerda. *Trans. from:* Compt. rend. Acad. d. sc., Par.] Gaz. med. da Bahia, 1881-2, 2. S., VI, 549-559.

COUTY (L.). De l'action du permanganate de potasse contre les accidents du venin des Bothrops. Compt. rend. Acad. d. sc., Par., 1882, XCIV, 1198-1201.

COUTY et DE LACERDA. Sur l'action du venin du bothrops jararacussu. Compt. rend. Acad. d. sc., Par., 1879, LXXXIX, 372-375. *Also:* Gaz. méd. de Par., 1879, 6. S., I, 460.

COUTY et DE LACERDA. Sur la difficulté d'absorption et les effets locaux du venin du bothrops jaracaca. Comp. rend. Acad. d. sc., Par., 1880, XCI, 549-551.

COUTY et DE LACERDA. Sur la nature inflammatoire des lésions produites par le venin du serpent bothrops. Compt. rend. Acad. d. sc., Par., 1881, XCII, 468-470.

CREUTZER (L.). Krankheitserscheinungen nach einem Schlangenbisse (der vipera chersea); an sich selbst beobachtet und mitgetheilt. Ztschr. d. k. k. Gesellsch. d. Aerzte zu Wien, 1853, I, 40-45.

CRUSSARD. Morsure de vipère suivie d'accidents généraux graves; amélioration rapide après la cautérisation au fer rouge. Ann. Soc. méd. de l'arrond. de Neufchâteau, 1874, I, 19. *Also:* Rev. méd. de l'est, Nancy, 1876, V, 247.

CURTIS (D. G.). Trigonocephalus con. poison (copperhead). Am. Homœop., N. Y., 1879, V, 293.

CZERMAK (J. J.). Beobachtungen über den Biss giftiger Schlangen. Verhandl. d. k. k. Gesellsch. d. Aerzte zu Wien, 1842, I, 337-343.

D. (R.). Dr. Shortt and Professor Gautier on snake poison. Med. Times and Gaz., Lond., 1881, II, 373.

DANIELLI (D.). Sei veneficii ipostenici curati giusta i precetti della dottrina medica italiana. Gazz. med. ital., prov. venete, Padova, 1860, III, 125.

DARELIUS (J. A.). Ligaum colubrinum. Diss. inaug. 8°. Upsaliæ, 1749. [Linn. Diss., Vol. II.]

DARLINGTON (W.). Remarks on Dr. Hugh Williamson's opinions concerning the fascination of serpents. Med. Reposit., N. Y., 1808, V, 257-260.

DARWIN (R. T.). Snake bite; recovery. Indian M. Gaz., Calcutta, 1882, XVII, 159.

DAY (F.). On the bite of the sea snake. Indian M. Gaz., Calcutta, 1869, IV, 92.

Death from rattlesnake bite. Med. and Surg. Reporter, Phila., 1869, XXI, 93.

DECHILLY. Morsure de vipère; cautérisation quatre heures après l'accident; guérison. Gaz. d. hôp., Par., 1831, V, 295.

DE-FAVERI (L.). Morso di vipera guarito cogli alcoolici. Gazz. med. ital., prov. venete, Padova, 1862, V, 285.

DE LA ROCHE BRAUNE (J.). On a case of snake bite. Austral. M. J., Melbourne, 1864, IX, 97-99.

DELASIAUVE. Morsure de vipère suivie de guérison. Gaz. d. hôp., Par. 1872, XLV, 874.

DELASIAUVE. Sur la morsure de la vipère. Union méd., Par., 1878, XXV, 107.

DELPECH. Sur les morsures de vipères. Bull. Acad. de méd., Par., 1874, XXXVIII, 608-610.

DEMEURAT (L.). Observation d'accidents développés à la suite d'une morsure de vipère, et se reproduisant depuis trente-neuf ans d'une manière parfaitement périodique. Gaz. hebd. de méd., Par.,1863, X, 736.

DEMPSTER. Case of death from the bite of a poisonous snake (Bungarus lineatus, Daudin). Edinb. M. and S. J., 1837, XLVII, 165.

DESAULX. Relation de quelques nouvelles expériences faites avec le venin de la vipère. Arch. gén. de méd., Par., 1827, XIII, 518-522.

DESGRANGES. Observations sur l'emploi de quelques plantes fraîches contre les morsures des vipères. Hist. Soc. de méd, prat. de Montpel., 1807, XVI, 265-276.

D'SOUZA (V.). On the treatment of snake bites. Indian Lancet, Lahore, 1859, I, 119.

D'SOUZA (V.). Natural history of some of the indigenous snakes of Scinde. Tr. M. and Phys. Soc. Bombay, 1869, N S., IX, 37-47, 15 pl.

DEVADE. Emploi de la bardane contre la morsure de la vipère. J. de méd. et chir. prat., Par., 1869, 2. S., XI, 355.

DIXON (G. F.). Case of snake bite. Austral. M. J., Melbourne, 1860, V, 112.

DRIOUT. Rapport au Conseil de santé des armées sur le traitement des morsures de vipères à cornes. Rec. de mém. de méd. . . . mil., Par., 1882, 3. S., XXXVIII, 420-424.

DRIOUT. Le permanganate de potasse dans le traitement de la morsure des vipères à cornes. Gaz. méd. de l'Algérie, Alger, 1883, XXVIII, 59.

DUDON (E.). Morsure de vipère; injection intraveineuse d'ammoniaque, guérison. Bordeaux méd., 1874, III, 117-119.

Effets incroyables d'une morsure de vipère. Gaz. méd. de Lyon, 1867, XIX, 422.

EGEBERG (C. A.). Död efter Bid af Coluber berus hos et 4 aar gammelt barn. Norsk Mag. f. Lægevidensk., Christiania, 1861, XV, 41-47.

EHRENREICH. Schlangenbiss. Med. Ztg., Berl., 1842, XI, 53.

VAN EIJK (J. A.). Over het slangenvergift en de geneesmiddelen daartegen aangewend. Volksvlijt (De), Amst., 1882, Nos. 9-10, 257-275.

ELDER (T. A.). Ammonia in snake bite. Chicago M. J., 1872, XXIX, 740-742. *Also*: Med. and Surg. Reporter, Phila., 1877, XXXVII, 118.

EMANUEL (L.). Fatal case of snake bite. Med. Times and Gaz., Lond., 1863, I, 449.

EMERY (C.). Ueber den feineren Bau der Giftdrüse der Naja haje. Arch. f. mikr. Anat. Bonn, 1875, XI, 561-568, 1 pl.

ENCOGNÈRE (JACQUES). Des accidents causés par la piqûre du serpent de la Martinique et de leur traitement. Thèse. 4°. Montpellier, 1865, No. 7.

Enquête sur les morsures de vipères dans le département de la Loire, faite en janvier et février, 1862, par la Société de médecine. Ann. Soc. de méd. de St.-Étienne et de la Loire, 1865, II, 392-408.

ETESON (A.). The snake stone. Indian M. Gaz., Calcutta, 1876, XI, 309-313.

EWART (J.). How the bite of snakes, supposed to be poisonous, may be cured. Indian M. Gaz., Calcutta, 1871, VI, 232.

Experiments at St. Bartholomew's Hospital to test the effect of Mr. Higgins's antidote to snake poison. Rep. on san. meas. in India, 1873-4. Lond., 1875, VII, 274.

Experiments on the bite of snakes, and the degree of danger in them. Med., Chir. and Anat. Cases. 8°. Lond., 1758, 61-68.

FAITHORN (J.). An account of the bite of a viper, cured by applying the fat of the same animal. New Lond. M. J., Lond., 1792, I, 345-347.

FALLE von Vipernbiss. Mag. f. d. ges. Heilk., Berl., 1826, XXI, 240-243.

FANEAU DE LA COUR (É.). Mémoire sur les effets de la morsure d'un lézard comparés à ceux de la morsure de la vipère. J. univ. d. sc. méd., Par., 1824, XXXIV, 5-18.

FANEAU DE LA COUR (É.). Deuxième mémoire sur la morsure de la vipère, pour faire suite au premier inséré dans le cahier d'avril, 1824, du Journ. univ. d. sc. méd. J. univ. d. sc. méd., Par., 1826, XLIII, 5-40.

FANEAU DE LA COUR (É.). Observations sur la morsure des serpens. J. univ. d. sc. méd., Par., 1829, LIV, 35-57.

FANEAU DE LA COUR (É.). Mémoire sur la morsure des ophidiens. Ann. de la méd. phys., Par., 1833, XXIII, 273-307; 399-405.

FAYRER (J.). Experiments on the influence of snake poison. Indian M. Gaz., Calcutta, 1868, III, 169; 193; 217; 241; 265; 1869, IV, 3.

FAYRER (J.). On the Cobra poison. Edinb. M. J., 1868-9, XIV, 522, 915, 996; 1869-70, XV, 236, 334, 417, 620, 805, 994, 1099; 1870-71, XVI, 48, 135, 237, 320, 423, 623, 715, 1101.

FAYRER (J.). Experiments on the influence of certain reputed antidotes for snake poison. Indian M. Gaz., Calcutta, 1869, IV, 25.

FAYRER (J.). Experiments on the influence of snake poison and on the injection of liquor ammoniæ into the venous circulation as an antidote. Indian M. Gaz., Calcutta, 1869, IV, 129-132.

FAYRER (J.). Experiments on the influence of snake poison and on the injection of certain fluids into the venous circulation, as antidotes, and on the application of the ligature and actual cautery. Indian M. Gaz., Calcutta, 1869, IV, 153-156.

FAYRER (J.). Deaths from snake bites; a trial condensed from the sessions' report. Indian M. Gaz., Calcutta, 1869, IV, 156.

FAYRER (J.). Experiments on the influence of the poison of the cobra, the daboia, and the bungarus, and of certain methods of treatment. Indian M. Gaz., Calcutta, 1869, IV, 177.

FAYRER (J.). Experiments on the influence of snake poison, and the use of certain reputed antidotes; and the effects of excision, etc. Indian M. Gaz., Calcutta, 1869, IV, 201-204.

FAYRER (J.). Experiments on the influence of snake poison, and on the effects of certain methods of treatment. Indian M. Gaz., Calcutta, 1869, IV, 225-228.

FAYRER (J.). The Thanatophidia of India, being a description of the venomous snakes of the Indian peninsula, with an account of the influence of their poison on life; and a series of experiments. X, 156 pp., 31 pl. fol. London, J. & A. Churchill, 1872.

FAYRER (J.). Experiments on the influence of snake poison on the blood of animals. Indian M. Gaz., Calcutta, 1869, IV, 249.

FAYRER (J.). The Thanatophidia of India. Indian M. Gaz., Calcutta, 1870, V, 1, 25, 49, 97, 117, 137, 157, 177, 197, 217, 237; 1871, VI, 1, 21.

FAYRER (J.). Experiments on snake poison. Indian M. Gaz., Calcutta, 1870, V, 119, 140, 158, 198, 219, 240.

FAYRER (J.). Notes on deaths from snake bite in the Burdwan Division. Indian Ann. M. Sc., Calcutta, 1870, V, 163-175.

FAYRER (J.). On the influence of the poison of bungarus cœruleus or Krait. Indian M. Gaz., Calcutta, 1870, V, 181.

FAYRER (J.). Experiments on snake poison. Indian M. Gaz., Calcutta, 1870, V, 182, 198, 219, 240; 1871, VI, 3, 24.

FAYRER (J.). On the immediate treatment of persons bitten by venomous snakes. Indian M. Gaz., Calcutta, 1871, VI, 26.

FAYRER (J.). Another antidote for snake poison. Indian M. Gaz., Calcutta, 1871, VI, 174.

FAYRER (J.). Fatal case of snake bite. Indian M. Gaz., Calcutta, 1871, VI, 175.

FAYRER (J.). Case of snake bite. Indian M. Gaz., Calcutta, 1872, VII, 11.

FAYRER (J.). Treatment of snake poisoning by artificial respiration. Indian M. Gaz., Calcutta, 1872, VII, 218.

FAYRER (J.). Experiments on cobra poison and on a reputed antidote. Indian M. Gaz., Calcutta, 1873, VIII, 6.

FAYRER (J.). Experiments on the poison of the rattlesnake. Med. Times and Gaz., Lond., 1873, I, 311.

FAYRER (J.). Snake poisoning in India. Med. Times and Gaz., Lond., 1873, II, 249, 492.

FAYRER (J.). The ammonia treatment of snake poisoning. Med. Times and Gaz., Lond., 1874, I, 601.

FERRARIO (E.). Morsicatura d'una vipera; uso interno dell' ammoniaca col laudano; guarigione. Gazz. med. Ital. lomb., Milano, 1852, 3. S., III, 422.

Sic FERRIER (J. ALPHONSE). Des morsures de vipère et de leur traitement. Thèse. 50 pp. 4°. Paris, 1858, No. 143.

FEUVRIER (J. B.). Deux cas de morsure de serpent venimeux; injection intra veineuse d'ammoniaque dans un cas. 8°. Paris, 1874. Not a reprint—see also infra.

FEUVRIER. Deux cas de morsure de serpent venimeux; injection intra-veineuse d'ammoniaque dans un cas. Gaz. hebd. de med., Par., 1874, XXI, 416.

Few (A) words on snakes; visit to Dr. Shortt; experiments on different snake poisons and different animals; properties of potass as an antidote. Med. Times and Gaz., Lond., 1873, II, 213.

First and second instalments of a report by the committee appointed in Calcutta by the Government of India to carry out in Bengal the suggestions made by Dr. Fayrer and Dr. Brunton as to the use of artificial respiration in snake bite. Rep. on . . . san. improve. in India 1872-3, Lond., 1873, 258-264.

FISCHER. Ein Biss der Coluber berus L. Ztschr. f. Wundärzte u. Geburtsh., Stuttg., 1866, XIX, 7.

FLETCHER (ROBERT). A study of some recent experiments in serpent venom. 16 pp. 8°. [Philadelphia], 1883. Repr. from: Am. J. M. Sc., Phila., 1883.

FLETCHER (R.). A study of some recent experiments on serpent venom. Am. J. M. Sc., Phila., 1883, N. S., LXXXVI, 131-146.

FONTANA. Treatment of snake bite by intravenous injections of ammonia a century ago. Med. Times and Gaz., Lond., 1873, II, 216.

FOOT (M.). An examination of Dr. Hugh Williamson's memoir on fascination; to which is subjoined a new theory of that phenomenon. Med. Reposit., N. Y., 1808, V, 113-122.

FRANCIS (C. R.). On the action of cobra poison. Indian Ann. M. Sc., Calcutta, 1868, No. XXIV, 297-317, 1 chart. Also Abstr.: Indian M. Gaz., Calcutta, 1868, III, 73.

FRANCIS (C. R.). On snake poison. Indian M. Gaz., Calcutta, 1868, III, 125.

FRANCIS (C. R.). Experiments on snake bite. Med. Times and Gaz., Lond., 1874, II, 259.

FRANK (B.). Chlor gegen Viperngift. Wchnschr. f. d. ges. Heilk., Berl., 1847, 527–532.

FREDET. Cas de mort par la morsure d'une vipère. Gaz. d. hôp., Par., 1872, XLV, 841–843.

FREDET. Considérations sur la morsure de la vipère, en Auvergne. Assoc. franç. pour l'avance. d. sc. Comp.-rend. 1876, Par., 1877, V, 817–827. *Also:* Union méd., Par., 1878, XXV, 74, 85.

GAUTIER (A.). Sur le venin du naja tripudians (cobra capello) de l'Inde. Bull. Acad de méd., Par., 1881, 2. s., X, 947–958.

GEIS. Vipernbiss. Med. Ztg., Berl., 1836, V, 105.

GEMELLI (S.). Di un avvelenamento per morso di vipera, e d'altri. Gazz. med. ital., prov. venete, Padova, 1864, VII, 410–413.

GERRARD (J.). Snake poison and its alleged antidotes. Austral. M. Gaz., Melbourne, 1870, II, 27.

GHIDELLA (P.). Storia di morsicatura di vipera. Gior. di chir.-prat., Trento, 1828, VI, 266–269.

GILLESPIE. [On the bites of serpents.] Med. and Phys. J., Lond., 1800, IV, 293–295.

GISTL (J.). Schlangen des Alterthums; ihr Gift und dessen Gegenmittel. Med. Centr.-Ztg., Berl., 1832, I, 529, 545, 561.

GLEEDE. Tödtlicher Schlangenbiss. Wchnschr. f. d. ges. Heilk., Berl., 1843, 66.

GRAND-BOULOGNE (A.). Morsure de la vipère. Nouveaux détails sur l'organisation de ce reptile. J. d. conn. méd.-chir., Par., 1839, 188–191.

GRANT (W. T.). The rattlesnake's poison, and its remedies. Georgia M. Companion, Atlanta, 1871, I, 457–459.

GRANT (W. T.). [Cnicus benedictus, cure for snake bite.] Atlanta M. and S. J., 1878, XV, 585–590.

GROSS (F.). De l'expectation dans le traitement des morsures de la vipère indigène. Rev. med. de l'est, Nancy, 1875, II, 317–324.

Guérison d'une morsure de vipère, opérée par l'eau de Luce. [Mém. Acad. r. d. sc., 1766.] Collect. acad. d. mém., etc., Par., 1787, XIV, 306.

GUNNING (J. D.). Case of snake bite; recovery. Indian M. Gaz., Calcutta, 1882, XVII, 294. *Also:* Brit. M. J., Lond., 1882, II, 888.

GUNNING (J. W.). Case of snake bite; recovery. Med. Times and Gaz., Lond., 1875, II, 518.

GUYON. Morsure du céraste ou vipère cornue. Gaz. méd. de Par., 1862, 3. s., XVII, 71–73.

HALFORD (GEORGE B.). On the condition of the blood after death from snake bite, as a probable clue to the further study of zymotic diseases, and of cholera especially. 24 pp. 8°. Melbourne, Stillwell & Knight, prtrs., 1867.

HALFORD (G. B.) The new treatment of snake bite, with plain directions for injecting. 8°. Melbourne, 1869.

HALFORD (G. B.). The treatment of snake bite in Victoria. 8°. Melbourne, 1870. *Repr. from:* Austral. M. J., Melbourne, 1870, XV.

HALFORD (G. B.). On the injection of ammonia into the veins. 7 pp. 8° [Melbourne, 1871?.]

HALFORD (G. B.). The poison of the cobra. Austral. M. J., Melbourne, 1867, XII, 153–155.

HALFORD (G. B.). Experiments on the poison of the cobra-di-Capello Brit. M. J., Lond., 1867, II, 43.

HALFORD (G. B.). Further observations on the condition of the blood after death from snake bite. Brit. M. J., Lond., 1867, II, 563.

HALFORD (G. B.). Tabular list of cases of snake bite treated by the injection of liquor ammoniæ. Austral. M. J., Melbourne, 1870, XV, 5.

HALFORD (G. B.). The treatment of snake bite in Victoria. Austral. M. J., Melbourne, 1870, XV, 161–176.

HALFORD (G. B.). Snake poisoning and its treatment. Med. Times and Gaz., Lond., 1873, II, 90, 170, 224, 323, 461, 575, 712.

HALFORD (G. B.). The treatment of snake poisoning. Med. Times and Gaz., Lond., 1874, I, 53.

HALFORD (G. B.). Du traitement des morsures de serpents venimeux par les injections intra-veineuses d'ammoniaque. Bull. gén de thérap., etc., Par., 1874, LXXXVII, 258–271.

HALFORD (G. B.). On the effects of the injection of ammonia into the veins in cases of snake poisoning [and discussion]. Austral. M. J., Melbourne, 1875, XX, 66-135.

HALL (E. H.). Prof Halford's treatment of snake bite by the injection of liquid ammonia into the veins. Calif. M. Gaz., San Fran., 1869-70, II, 229.

HALLOWELL (E.). Remarks on the bites of venomous serpents, with cases. Tr. Coll. Phys, Phila., 1850, N. S., I, 394-401.

HAM (L. S.). Case of rattlesnake bite. Buffalo M. and S. J. and Reporter, 1861, I, 82-85.

HANCOCK (J.). Treatment of snake bites. Lancet, Lond., 1830, I, 800-802.

HANCOCK (J.). Remarks on snake poisons and their remedies. Med. Times, Lond., 1839, I, 132.

HANKEL (E.). Zwei Fälle von Vergiftungen durch Otterbiss. Arch. d. Heilk., Leipz., 1876, XVII, 264-267.

HANLEY (W.). On snake bites. North-West. M. and S. J., Chicago, 1854, XI, 497-499.

HARDISON (W. H.). Ammonia in the treatment of bites of poisonous reptiles and insects. Louisville M. News, 1880, IX, 270.

HARLAN (R.). Experiments made on the poison of the rattlesnake; in which the powers of the Hieracium venosum, as a specific, were tested; together with some anatomical observations on this animal. Tr. Am. Phil. Soc, Phila., 1828, N. S., III, 300-314.

HARWOOD (E.). Iodine, an antidote to the poison of the rattlesnake. Northwest M. and S. J., Chicago, 1854, N. S., XI, 187.

VAN HASSELT (A. W. M.). Der Schlangenbecher von Ceylon. Arch. f. d. Holländ. Beitr. z. Nat.- u. Heilk, Utrecht, 1861-4, III, 222.

HAUSSMANN (ALBERT-JUL.). De morsu serpentum. Diss. inaug. 29 pp. 8°. Regiomonti Prussorum, impr. J. Dalkowski, 1859 (?)

HAUX. Ein Fall von Schlangenbiss. Med. Cor.-Bl. d. württemb. ärztl. Ver., Stuttg., 1841, XI, 226.

HAYNES (J. R.). Experiments in animal poisons. Crotalus horridus (rattlesnake). Cincio. M. Advance. 1879, VI, 481-487.

HAYNESWORTH. Some account of the success of the plant called Jestis-weed, in curing the disease induced by the bites of the rattlesnake, and other venomous serpents. Phila. M. and Phys. J., 1808, III, 57-61.

HEDLEY and REES. Fatal case of snake bite, presenting some curious points of interest. Austral. M. J., Melbourne, 1874, XIX, 49-53.

HEIDENSTAM (F. C.). Case of snake bite. Lancet, Lond., 1881, I, 290.

HEINZEL (L.). Zur Pathologie des Vipernbisses. Wien. med. Presse, 1865, VI, 1172.

HEINZEL (L.). Ueber einen Fall von Schlangenbiss. Wien. med. Presse, 1865, VI, 707.

HEINZEL (L.). Zur Pathologie und Therapie der Vergiftung durch Vipernbiss. Wchnbl. d. Ztschr. d. k. k. Gesellsch. d. Aerzte in Wien, 1866, XXII, 169, 181, 193, 205, 217, 229, 240. Also, abstr.: Wien. med. Presse, 1866, VII, 47-49. Also, abstr.: Wien. med. Wchnschr., 1866, 7.

HEINZEL (L.). Ein Todesfall durch Vipernbiss. Wchnbl. d. k. k. Gesellsch. d. Aerzte in Wien, 1867, VII, 121, 133.

HELM. Folgen eines Natternbisses. Mag. f. d. ges. Heilk., Berl., 1827, XXIII, 311.

HÉMARD. Gangrène momifiante du doigt indicateur par suite d'une morsure de vipère. Bull. Soc. de chir. de Par., 1859, IX, 377-379.

HENDERSON (C.). Case of snake bite. Austral. M. J., Melbourne, 1880, N. S., II, 64.

HENRY. Sur un cas de morsure de vipère. Rap. de M. Merat. Bull. Acad. de med., Par., 1843-4, IX, 1021-1023.

HENSINGER. Ueber die Wirkungen des Klapperschlangenbisses. Mag. f. d. ges. Heilk., Berl., 1822, XII, 443-448.

HILL (J. J.). Case of snake bite cured by the internal administration of ammonia. N. South Wales M. Gaz., Sydney, 1870-1, I, 207.

HILL (J. J.). A case of successful treatment of snake bite. N. South Wales M. Gaz., Sydney, 1873-4, IV, 163.

HILSON (A. H.). Two cases of snake bite, treated by injection of liquor ammoniæ into the veins, with remarks on the action of snake poison. Rep. on . . . san. improve. in India, 1872-3, Lond., 1873, 283-288. Also: Indian M. Gaz., Calcutta, 1873, VIII, 258-260.

HOIT (M.). Snake bites. Boston M. and S. J., 1844, XXX, 243.

HOOD. On the treatment of poisoning by the cobra. Lancet, Lond., 1868, I, 221.

HOPKINS (W. K.). Alcohol as a remedy for the poison of the rattlesnake. Northwest. M. and S. J., Chicago, 1852-3, IX, 389-391.

HOPPE-SEYLER (F.). Ueber die Zusammensetzung der Blutkörperchen des Igel und Coluber natrix. Med.-chem. Untersuch. a. d. Lab., . . . zu Tübing., Berl., 1868, I, 391-393.

HUGHES (A. H.). Case of recovery from the bite of a cobra. Tr. M. and Phys. Soc. Bombay, 1871, N. S., XI (App.), XXVII-XXIX.

HULIN-ORIGET. Observation sur une morsure de vipère. Rec. d. trav. Soc. méd. d'Indre-et-Loire, Tours, 1841, 38-42.

HUNTER. [Case of snake bite.] Tr. M. and Phys. Soc. Bombay, 1842, V, 54.

HUSANT. Morsure de vipère traitée par le suc d'euphorbe. J. de chim. méd., etc., Par., 1839, 2. s., V, 272-276.

HUSSA (F.). Tödtlicher Schlangenbiss. Allg. Wien. med. Ztg., 1861, VI, 264.

IMLACH (C. J. F.). Mortality from snake bites in the province of Sind; from official records, with a special report on the snake season of 1854. Tr. M. and Phys. Soc. Bombay (1855-6), 1857, N. S., III, 80-130.

Important case of snake bite at Albury, N. S. W. Austral. M. Gaz., Melbourne, 1870, II, 26.

INGALLS (W.). Bite of a moccasin snake. Boston M. and S. J., 1843. XXVII, 170.

Intravenous injection of ammonia in snake bite. Med. Times and Gaz., Lond., 1876, II, 464-466.

IRVING (J.). The negro Cæsar's cures for poison, and the bite of a rattlesnake. For discovering of which the Gen. Assembly of North Carolina thought fit to purchase his freedom, and grant him an allowance of £100 per annum during life. Gentleman's Mag. and Hist. Chron., 1750, XX, 342. [*From:* Carolina Gaz., May, 1750.]

IRWIN (B. J. D.). Notes on "Euphorbia prostrata" as an antidote to the poison of the rattlesnake. Am. J. M. Sc., Phila., 1861, N. S., XLI, 89-91.

JACKSON (M. H.). Rattlesnake bites. South. Pract., Nashville, 1879, I, 259, 360.

JACKSON (S.). Remarks on the bites of venomous serpents, with cases. Tr. Coll. Phys. Phila., 1850, N. S., I, 402-411.

JACOBI PÈRE. Morsure de vipère. Mém. Soc. de méd. de Strasb., 1870, VII, 298-300.

JACOLOT. Die Curados de Culebras, oder Impfung zum Schutze gegen den Biss giftiger Schlangen; ein auf authentischer Forschung in Mexico basirter Originalbericht. Wien. med. Wchnschr., 1867, 731, 747.

JACQUEMET. De la morsure de la vipère Naja, en Algérie, et de son traitement par l'acide phénique. Rec. de mém. de méd. . . . mil., Par., 1881, 3. s., XXXVII, 226-231.

JAMES (G. W.). Case of viper bite. Lancet, Lond., 1832, I, 294.

JAMES (R. W.). Case of snake bite. Tr. M. and Phys. Soc. Bombay (1857-8), 1859, N. S., IV, app. xxxix.

JARJAVAY. Morsure de vipère. Gaz. d. hôp., Par., 1861, XXXIV, 358.

JEANBERNAT (ERNEST). Des animaux venimeux de la France. Thèse. 54 pp. 4°. Paris, 1862, No. 186.

JENKINS (G. W.). Observations on the pathology and treatment of the bite of the rattlesnake. Tr. Wisconsin M. Soc., Milwaukee, 1878, XII, 63-65.

JENKINS (W. H.). Case of snake bite; injection of ammonia into veins of both arms; recovery. Lancet, Lond., 1873, I, 900.

JOHNSON (W. D.). Treatment of snake bite. N. Orl. M. and S. J., 1860, XVII, 487.

JOLLEY (W. H.). Case of snake bite. South. Pract., Nashville, 1882, IV, 204.

JONES (G. W.). Bite of an adder. Lancet, Lond., 1828, II, 78-80.

JONES (J. T.). Rattlesnake bite. South. J. M. and Phys. Sc., Knoxville, 1857, VI, 376.

JÖRS. Vipern-Biss. Med. Ztg., Berl., 1855, 145.

JOSSO. Morsure de vipère; accidents graves; emploi du jaborandi; guérison. Gaz. hebd. de méd., Par., 1882, 2. S., XIX, 835.

KABIERSKE (C.). Beobachtung und Heilung einer Lymphangitis und Narcosishöhern

Grudes in Folge eines Natternbisses. Allg. med. Centr.-Zig., Berl., 1860, XXIX, 89.

KAISER (C. L.). Beobachtung über die Folgen eines Bisses der gemeinen Otter (Coluber berus). Heidelb. klin. Ann., 1832, VIII, 323.

KEARSLEY. Extract from a letter to Mr. P. Collinson on the bite of the rattlesnake, and Indian method of cure. Gentlem.'s Mag., Lond., 1766, XXXVI, 73-76.

KEITH (W.). Inhalation of ammonia as an antidote to snake poison. Kansas City M. J., 1875, III, 337-340.

KELP. Tod durch Schlangenbiss. Vrtljschr. f. gerichtl. Med., Berl., 1872, XVI, 98-101.

KELP. Ueber Schlangengift und die Kreuzotter. Friedreich's Bl. f. gerichtl. Med., Nürnb., 1883, XXXIV, 194-199.

KESTEVEN (W. B.). Is arsenic eating prophylactic against the effects of bites of venomous reptiles? Brit. M. J., Lond., 1858, 174.

KEY (B. P.). (Case of snake bite.) Nashville J. M. and S., 1877, XIX, 129.

KING (E. P.). Bite of a copperhead successfully treated by indigo. Annalist, N. Y., 1848, II, 229.

KLEINSCHMIDT. Zur Behandlung des Schlangenbisses durch subcutane Injection von Liquor ammon. caust. Berl. klin. Wchnschr., 1874, XI, 286.

KLEINSCHMIDT. Case of bite by a copperhead snake. Tr. M. Soc. Dist. Columbia, 1875, II, 54-56.

KLINE (L. B.). Case of septic poisoning caused by the bite of a copperhead. Med. and Surg. Reporter, Phila., 1868, XIX, 326.

KNOTT (J. J.). Case of snake bite successfully treated with calomel and iodide of potassium. Confed. States M. and S. J., Richmond, 1864, I, 213.

KNOTT (J. J.). A case of rattlesnake bite successfully treated by injections of carbonate of ammonia into veins. Med. and Surg. Reporter, Phila., 1877, XXXVII, 46-48.

KNOTT (J. J.). Injections of carbonate of ammonia for snake bite. Med. and Surg. Reporter, Phila., 1877, XXXVII, 79.

KNOX. On the treatment of wounds inflicted by poisonous snakes. Lancet, Lond., 1839, I, 199-203.

KOEPPEN. Heilungsgeschichte einer merkwürdigen, durch einen Biss von einer Schlange in den Hodensack bewirkten Verwundung. J. d. Chir. u. Augenh., Berl., 1834, XXI, 291-294.

KOEPPEN. Verletzung des Hodensacks durch einen Schlangenbiss und freiwillige Castration. Wchnschr. f. d. ges. Heilk., Berl., 1836, 221.

KOPPITZ (W.). Kreuzotterbiss bei einer Kuh. Oesterr. Monatschr. f. Thierh., Wien, 1879, IV, 81-83.

KÜCHENMEISTER (F.). Die Bisse der Vipera berus [d. i. der gemeinen Viper oder Kreuz-oder Haselotter]. Allg. Ztschr. f. Epidemiol., Stuttg., 1876, II, 345-370.

KUHN. Iets over de, in Suriname onder den naam van Oroekoekoe snecki bekende slang, beurens eene waarneming van derzelver beet. Ippocrates, Rotterd., 1819, V, 318-340.

KUNKLER (G. A.). On the bite of a copper snake. Med. Counselor, Columbus, 1855, I, 481-483.

KUNKLER (G. A.). Case of snake bite; death. Cincin. Lancet and Observ., 1859, N. S., II, 662.

LABORDE. Observation sur la guérison d'un chien, par le moyen de l'eau de Luce, qui, depuis soixante heures, avoit été mordu par un serpent à sonnettes. J. de méd., chir., pharm., etc., Par., 1774, XLI, 534-539.

DE LACERDA (J. B.). Venin des serpents. Compt. rend. Acad. d. sc., Par., 1878, LXXXVII, 1093-1095.

DE LACERDA (J. B.) O permanganato de potassa como antidoto da peçonha dos cobras. Uniao med., Rio de Jan., 1881, I, 514-519; 561-568. Also: Gaz. med. da Bahia, 1881-2, 2. S., VI, 153-158; 191-197.

DE LACERDA (J. B.). Sur le permanganate de potasse employé comme antidote du venin de serpent. Compt. rend. Acad. d. sc., Par., 1881, XCIII, 466-468.

DE LACERDA (J. B.). A acçao do alcool e do chloral sobre o veneno ophidico. Uniao med., Rio de Jan., 1882, II, 76-83, 109-116.

LACOMBE (A.). Treatment of the bites of poisonous reptiles. Boston M. and S. J., 1851, XLIV, 119.

LACOMBE (A.). Bites of a rattle and other poisonous snakes treated in Venezuela. Boston M. and S. J., 1851, XLIV, 289-292.

LALLEMANT (C.). Erpetologie de l'Algérie. Gaz. méd. de l'Algérie, Alger, 1866, XI, 19, 31, 47, 64, 81, 99, 128.

LAMBERT (G.). Poisoning by the bite of an adder. Lancet, Lond., 1851, II, 437.

LANDERER (X.) Des moyens usités en Orient pour se garantir des serpents et de se guerir de leurs morsures. Echo med., Neuchâtel, 1861, V, 498.

LANE (J. P.). A case of snake poisoning. Austral. M. J., Melbourne, 1870, XV, 58-60.

LANKESTER (E.). Bites of poisonous snakes and remedies. Lancet, Lond., 1839, I, 497-500.

LANSZWEERT (L.). Arseniate of strychnia; a new antidote to the poison of snakes. Pacific M. and S. J., San. Fran., 1871-2, V, 108-115.

LARREY. Analyse de deux observations de morsures de vipère. Compt. rend. Acad. d. sc., Par., 1874, LXXVIII, 1793.

LEBEL. Observation sur le traitement d'un soldat mordu par une vipère. J. de med mil., Par., 1782, I, 126-136.

LE CARPENTIER (J.). Case of snake bite complicated with a peculiar artificial constriction of the penis. Med. Press. and Circ., Lond., 1869, VIII, 380.

LECOMTE. Note sur les vipères du Sud d'Algérie. Gaz. med. de l'Algérie, Alger, 1870, XV, 4-6.

LENZ (H. O.). Biss der Viper Teutschlands und dessen Folgen bei Menschen und Thieren. J. d. pract. Heilk., Berl., 1830, LXXI, 4. St., 3-22.

LEROUX (C.). Glossite intense produite par une piqûre de vipère. Ann. d. mal. de l'oreille et du larynx, Par., 1878, IV, 258-261.

LE ROY DE MÉRICOURT. Du traitement des morsures des serpents venimeux de l'Inde et de l'Australie par l'ammoniaque. Bull gén. de thérap., etc., Par., 1874, LXXXVII, 362-365.

LE ROY DE MÉRICOURT. Discussion sur la morsure des serpents venimeux. Bull. Acad. de méd., Par., 1874, XXXVIII, 559-572.

Letter on snake poison. Lancet, Lond., 1871, I-599.

LEYDIG (F.) Zur Kenntniss der Sinnesorgane der Schlangen. Arch. f. mikr. Anat., Bonn, 1872, VIII, 317-357.

LINDSLEY (H.). Alcohol as a remedy for the poison of the rattlesnake. Stethoscope and Virg. M. Gaz., Richmond, 1852, II, 540.

LONG (A.). Case of death from snake bite. Indian M. Gaz., Calcutta, 1882, XVII, 294.

LOTOTZKY (J. F.). Case of snake bite. Madras Month. J. M. Sc., 1870, II, 428.

LOWNDS (T. M.). Two cases of snake bites. Tr. M. and Phys. Soc. Bombay, (1853-4), 1855, N. S., II, 328-330. *Also*: Month. J. M. Sc., Edinb., 1854, XVIII, 135-138.

LUGOL (F.). Le venin d'une seule vipère peut entrainer la mort. Bull. gén. de thérap., etc., Par., 1835, VIII, 86-88.

LUSTIG. Schlangenbiss. Med. Ztg., Berl., 1848, XVII, 206.

LUZ (A.). Mordedura de cobra cascavel; cura pela cauterisação e pelo oleo de recino, nitro e sudorificos internamente. Rev. med. do Rio de Jan., 1876, III, 200.

LYNSDALE (H. H.). Case of snake bite, treatment by cold douche, forced exercise, subcutaneous injection of liquor ammoniae, and galvanism; recovery. Indian M. Gaz., Calcutta, 1876, XI, 213.

LYONS (J. J.). A case of snake bite. N. Orl. M. and S. J., 1860, XVII, 526.

MACKENZIE (F. M.). Case of snake bite. Indian M. Gaz., Calcutta, 1872, VII, 134.

MACQUER. Guérison de la morsure d'une vipère, par l'eau de Luce. J. de méd, chir., pharm., etc., Par., 1766, XXV, 271-275.

MACRAE (J.). Case of the bite of a poisonous snake successfully treated. Med. and Phys. J., Lond., 1813, XXIX, 120-125.

McGIRR (J. E.). Notes of a rattlesnake bite, admitted into the Illinois Gen.-Hosp. of the Lakes. N. West, M. and S. J., 1851-2, VIII, 311-314.

McREDDIE (G. D.). A case of snake bite; treatment with permanganate of potash. Indian M. Gaz., Calcutta, 1882, XVII, 267.

MAITLAND. A case of snake bite. Proc. Hyderabad M. and Phys. Soc., Secunderabad, 1853, I, 128-130.

MANDIC. Zur Kasuistik des Vipernbisses. Wien. med. Presse, 1871, 640, 661, 683.

MANDT (C. C. W.). Doodelijke gevolgen van een slangenbeet. Tijdschr. d. Vereen. tot bevord. d. Geneesk, Batavia, 1857, V, 956-961.

MANGOLD. Ein tödtlicher Vipernbiss. Mag. f. d. ges. Heilk., Berl., 1825, XX, 155.

MARIANINI. Singular effect of the bite of a viper. Lancet, Lond., 1829, I, 580.

MARSCHALL (G.). Eine Verletzung von einer Viper, Coluber berus. Arch. f. med. Erfahrung, Berl., 1831, I, 159.

MARTIN (J. B.). Remarkable case of snake bite. St. Louis M. and S. J., 1859, XVII, 110-116.

MARTINEZ (M.). Discurso phisico sobre si la vivoras deban reputarse por carne, o pescado, en el sentido en que nuestra madre la Inglesia nos veda las carnes en dias de abstinencia? Repuesta a una consulta, que hicieron los R. R. Padres Cartujanos, para en vista de la resolucion poder usar las vivoras á lo menos como medicamento, lo qual en caso de reputarse por carne les feria vedada, segun su laudable costumbre. [In his: Noches anatomicas. 4°. Madrid, 1750, pp. 205-222.]

MASLIEURAT-LAGÉMARD (L.). De la morsure de vipère et de son traitement. 8°. Guéret, 1878. Repr. from: Écho de la Creuse, June 1, 1878.

MAURER. Giftige Wirkung des Vipernbisses. Mitth. d. badisch. ärztl. Ver., Karlsruhe, 1858, XII, 71.

MAYERUS (L. C.). Prodromus de medicamentorum viperinorum usu. 4°. n. p. (1694).

MAYO. Case of adder's bite. Lond. M. and S. J., 1832, I, 708.

MAZEIR. Morsure de vipère, traitée par les ventouses. Ann. de la méd. physiol., Par., 1828, XIV, 638-640.

MEANS (A.). On the modus operandi of the poison of venomous reptiles, etc. South. M. and S. J., Augusta, 1846, N. S., II, 1-11.

MEASE (J.). Remedies for the cure of diseases produced by the bites of snakes and mad dogs. Phila. M. Museum, 1808, V, 1-15.

MEDICI (G.). Due parole su la vipera; morte per morso viperino; sezione del cadavere. Gazz. med. ital., lomb., Milano, 1858, 4. s., III, 368-370.

MEISCHNER (E.). Wirkungen des Natternbisses. Allg. med Ann., Altenb., 1819, 803-807. Also transl. [Abstr.]: Med. and Phys. J., Lond., 1820, XLIII, 196.

METLER. Cure of snake bite by Dr. Halford's remedy (injection of ammonia). Med. Times and Gaz., Lond., 1870, I, 457.

MILLER (J.). On the effects of oil in cases of the bite of serpents. [From: Charleston, S. C., City Gazette.] Med. Reposit., N. Y., 1799, II, 253-255.

MIMRA (M.). Schlangenbiss. Allg. Wien. med. Ztg., 1865, X, 76.

MITCHELL (A.). Bite of the diamond rattlesnake (Crotalus adamanteus). Boston M. and S. J., 1873, LXXXIX, 331-333.

MITCHELL (S. W.). On the treatment of rattlesnake bite, with experimental criticisms upon the various remedies now in use. Am. M.-Chir. Rev., Phila., 1861, V, 269-310.

MITCHELL (S. WEIR). On the treatment of rattlesnake bites, with experimental criticisms upon the various remedies now in use. 45 pp. 8°. Philadelphia, J. B. Lippincott & Co., 1861. [Repr. from: N. Am. M.-Chir. Rev., March, 1861.]

MITCHELL (S. W.). Experimental contributions to the toxicology of rattlesnake venom. N. York M. J., 1867-8, VI, 289-322.

MITCHELL (S. WEIR). Experimental contributions to the toxicology of rattlesnake venom. 34 pp. 8°. New York, Moorhead, Simpson & Bond, 1868. Repr. from: N. York M. J.

MITCHELL (S. W.). The venom of serpents. Med. Times and Gaz., Lond., 1869, I, 137.

MITCHELL (S. W.). Observations on poisoning with rattlesnake venom. Am. J. M. Sc., Phila., 1870, Apr., 317-323.

MITCHELL (S. W.). Remarks upon some recent

investigations of the venom of serpents. Lancet, Lond., 1883, II, 94.

MITCHELL (S. W.) and REICHERT (E. T.). Preliminary report on the venoms of serpents. Med. News, Phila., 1883, XLII, 469–472.

MODESTIN. Fälle von Schlangenbiss; Heilung. Med.-chir. Centr.-Bl., Wien, 1876, XI, 373.

MOISEAU (A.). Sur les animaux venimeux du département de la Vendée, et sur le traitement à employer contre leurs morsures et figures. Thèse. 48 pp. 8°. Paris, An. XI.

MONSELICE (A.). Storia medica di una ragazza morsicata da una vipera. Gazz. med. ital. lomb., Milano, 1849, 2. S., II, 399.

MONTIN (L.). Remède contre la morsure des couleuvres venimeuses. [From: Mem. abreges de l'Acad. de Stockholm.] Collect. acad. d. mém., etc., Par., 1772, XI, 300.

MOODIE (J.). Case of a woman bitten by a viper. Med. and Phys. J., Lond., 1804, XI, 481–488.

MOORE (W. J.). A case of snake bite. Indian M. Gaz., Calcutta, 1868, III, 103.

MORAGNE (N. H.). Bite of a copperhead— *trigonocephalus contortrix*—treated with whiskey. South. M. and S. J., Augusta, 1853, N. S., IX, 81.

MORANDOTTI (P.). Morsicatura di una vipera uso interno dell' ammoniaca; morte. Gazz. med. ital. lomb., Milano, 1852, 3. S., III, 382.

MOREAU DE JOANNÈS (ALEXANDRE). Monographie du trigonocéphale des Antilles, ou grande vipère fer-de-lance de la Martinique. (Lue à l'Académie royale des sciences, dans sa séance du 5 août 1816) 42 pp. 8°. Paris, 1816. *Also:* J. de méd., chir., etc., Par., 1816, XXXVI, 326–365.

MOREHEAD (C.). Notes on the effects of the bite of the phoorsa snake. Tr. M. and Phys. Soc. Bombay (1849-50), 1851, No. X, Append. 304–308.

MORIARTY (T. B.). Treatment of snake bites. Med. Times and Gaz., Lond., 1863, II, 54.

Mort causée par la morsure de la vipère du Nord; observation de guérison par les applications d'alcali, les boissons sudorifiques et le débridement de la plaie. J. de méd. et chir. prat., Par., 1832, III, 263–265.

Mort par suite de la morsure du cobra di Capello. Union méd., Par., 1852, VI, 555.

DE MOURA (J. R.). Tratamento das mordeduras por cobras venenosas. Gaz. med. da Bahia, 1868, III, 87–90.

MUGNA (G. B.). Avvelenamento per morso di vipera. Gazz. med. ital, prov. venete, Padova, 1863, VI, 265.

MULVANY (J.). A Tamil woman far advanced in pregnancy bitten by a Tic Polonga; recovery of the mother and death of the child. Lancet, Lond., 1874, II, 446.

NAPHEGYI (G.). Treatment of rattlesnake bites by the Pinto Indians of Mexico. Med. and Surg. Reporter, Phila., 1868, XVIII, 249–252.

NEIDHART (C.). *Crotalus horridus:* its analogy to yellow fever, malignant, bilious, and remittent fevers. Demonstrated by the action of the venom on man and animals. With an account of Humboldt's prophylactic inoculation of the venom of a serpent, at Havana, Cuba. 2d ed. XVI, 9–87 pp. 8°. New York, W. Radde, 1868.

NETTER. Note sur un cas de morsure de vipère traitée par l'injection sous-cutanée d'ammoniaque et suivie de guérison. Rev. méd. de l'est, Nancy, 1874, II, 100.

NICATI. Note sur les morsures de serpents venimeux. Bull. Soc. med. de la Suisse Rom., Lausanne, 1870, IV, 383–386.

NICHOLSON (E.). On some popular errors regarding Indian snakes. Madras Month. J. M. Sc., 1870, II, 350–358.

NICHOLSON (E.). Statistics of deaths from snake bite. Brit. M. J., Lond., 1883, II, 448.

NICK. Vergiftung durch einen Schlangenbiss. Med. Cor.-Bl. d. württemb. ärztl. Ver., Stuttg., 1851, XXI, 169.

NICK. Verletzung durch Coluber-berus. Med. Cor.- Bl. d. württemb. ärztl. Ver., Stuttg., 1852, XXII, 219.

NITSCHKE (P.). Treatment of snake bites. Med. and Surg. Reporter, Phila., 1869, XX, 77.

OGLE (W.). Loss of speech from the bite of venomous snakes. St. George's Hosp. Rep., Lond., 1868, III, 167–176.

OKADE RIOYEI. [The treatment of snake bite.] Tokei Zasshi Osaka, Sept. 5, 1881.

OLIVERAS (E. J.). Whiskey as a neutralizing agent to the poison of the rattlesnake. Oglethorpe M. and S. J., Savannah, 1858-9, I, 224-228.

ORÉ. Injection d'ammoniaque dans les veines, pour combattre les accidents produits par la morsure de la vipère. Compt. rend. Acad. d. sc., Par., 1874, I, 983-985.

OTT. Ueber Schlangenbiss. Aerztl. Cor.-Bl. f. Böhmen, Prag, 1874, II, 11-16.

OTT (I.). Rattlesnake virus; its relations to alcohol, ammonia, and digitalis. Arch. Med., N. Y., 1882, VII, 134-141.

OTT (I.). The physiological action of the venom of the copperhead snake, *Trigonocephalus contortrix*. Virginia M. Month., Richmond, 1882-3, IX, 629-634.

OUSTALET (E.). Serpents. Dict. encycl. d. sc. méd., Par., 1880, 3. S., IX, 378-387.

OWEN (F.). A case of snake bite. N. Orl. M. and S. J., 1867, XX, 207-209.

OWEN (H. R.). Bite of a viper. Lond. M. Gaz., 1840, N. S., II, 337-341.

PAINE (A. G.). On rattlesnake poison, rabies canina, etc. South. M. Rec., Atlanta, 1875, V. 129-145.

PARRY (E. I.). On the poison of the adder. Lancet, Lond., 1862, I, 608.

PASCALIS (F.). A case of an extraordinary enlargement of the foot, leg, and thigh, effected by the bite of a snake; with physiological remarks on the quantity of blood which can be contained in the human body. Med. Reposit., N. Y., 1818, N. S., IV, 78-84, 1 pl.

PASQUIER (H.). Du pronostic et du traitement de l'envenimation ophidienne. Thèse. 4°. Paris, 1883.

PASSANO (P. A.). Études historiques, théoriques et pratiques sur quelques points relatifs aux morsures des serpents venimeux. Thèse. VIII, 9-110 pp. 3 l. 4°. Montpellier, 1880, No. 57.

PAUL (J. L.). Case of snake bite; recovery. Madras Month. J. M. Sc., 1871, IV, 102-104.

PAUL and SHORTT. Cases of snake bite. Med. Times and Gaz., Lond., 1873, II, 214.

PEAKE (H.). Antidotes to serpent poison. N. Orl. M. and S. J., 1860, XVII, 375-377.

PEDLER (A.). On cobra poison. Proc. Roy. Soc., Lond., 1878, XXVII, 17-29.

PEIGNEL (L. J.). Observation de morsure de vipère. Rec. de mém. de méd. . . . mil., Par., 1830, XXVIII, 301-305.

PEMBERTON (O.). Bite of the thumb by the common adder, followed by general prostration of the vital powers; swelling and extensive ecchymosis of the right upper extremity; recovery in twelve days. Lancet, Lond., 1849, II, 638.

PEMBERTON (O.). Bite of the face by the common adder, followed by vomiting, diarrhœa, and extreme general prostration. Reaction induced by large doses of ammonia and brandy; diffuse cellular inflammation; recovery. Lancet, Lond., 1851, II, 157.

PERACCA (M. G.). e DERIGINUS (C.). Esperienze fatte sul veleno del Colopeltis insignitus. Gior. d. r. Accad. di med. di Torino, 1883, 3. S., XXXI, 379-383.

PÉRUT. Observation d'un cas de morsure de vipère ayant entraîné la mort 60 jours après l'accident. Ann. Soc. méd. de l'arrond. de Neufchateau, 1874, I, 20. *Also:* Rev. méd. de l'Est, Nancy, 1876, V, 248.

PETITOT (B.). Observation sur la morsure d'un serpent au petit doigt du pied droit, suivie de sphacèle qui a nécessité l'amputation de la jambe. Gaz. méd. de Par., 1833, 2. S., I, 481. *Also:* Rec. de mém. de méd. . . . mil., Par., 1834, XXXVI, 224-230.

PIETERS (A.). Een gunstig afgeloopen geval van slangenbeet. Geneesk. Arch. v. d. Zeemacht, Nieuwediep, 1874, III, 163.

DE PIETRA SANTA. Antidote contre le venin du cobra. J. d'hyg., Par., 1881, VI, 373.

PIFFARD (H. G.). Periodical vesicular eruption following the bite of a rattlesnake. Med. Rec., N. Y., 1875, X, 62.

PIORRY. Considérations physiologiques sur la morsure d'une vipère, traitée avec succès par l'application de ventouses. Rev. méd franç. et étrang., Par., 1826, IV, 63-86.

PIQUET. Morsure de vipère. Ann. méd. physiol., Par., 1834, XXVI, 437.

PODOLSKI (A.). Ukaszenie przez żmije; objawy zatrucia; wyzdrowienie. [Bite of a viper; recovery.] Medycyna, Warszawa, 1882, X, 471–473.

POTTER (O. F.) and McELROY (Z. C.). Organic poisoning; a case of snake bite with some observations on the same. Buffalo M. and S. J., 1869–70, IX, 247–259.

POVALL (R.). An attempt to account for the origin of the belief in the "uncommon subtility," and "fascinating faculty," generally ascribed to the serpent. Phila. J. M. and Phys. Sc., 1824, VIII, 113–117.

POWELL (R. H.). Two fatal cases of snake bite; with remarks. Assoc. M. J., Lond., 1853, 773.

PRINGLE. Case of snake bite. Tr. M. and Phys. Soc., Calcutta, 1845, IX, 242–243.

PRINGLE (G. H.). Suggestions as to the nature and treatment of snake bite, and, incidentally, of cholera. N. South Wales M. Gaz., 1870–71, I, 291–294.

PRIOU. Observation sur une morsure grave de vipère suivie de réflexions sur les divers moyens employés jusqu'à ce jour contre cet accident. J. de la sect. de méd. Soc. Loire, Inf., Nantes, 1828, IV, 201–211.

Professor Halford's treatment of snake bite. Med Times and Gaz., Lond., 1869, I, 123, 227; 1871, II, 161.

PUTZ. Zur Behandlung des Schlangenbisses durch subcutane Injection von Liquor ammon. caust. Berl. klin. Wchnschr., 1873, X, 330.

QUAIN. Bite from the hooded snake—cobra de Capello—rapid death; autopsy. Lancet, Lond., 1852, II, 397–400.

DE QUATREFAGES. Note sur le permanganate de potasse, considéré comme antidote du venin des serpents, à propos d'une publication de M. J. B. de Lacerda. Compt. rend. Acad. d. sc., Par., 1882, XCIV, 488–490. *Also:* France méd., Paris, 1882, I, 319–321.

RAE (W.). On a case of snake bite treated by ammoniacal injection. Austral. M. J., Melbourne, 1869, XIV, 242–244.

RALPH (T. S.). Observations on the action of snake poison on the blood. Austral. M. J., Melbourne, 1867, XII, 353–361.

RANKEN (J.). An account of some experiments made upon dogs bitten by the cobra de Capello, or coluber naji. Edinb. M. and S. J., 1822, XVIII, 231–239.

RATH. Vergiftung eines Knaben durch den Biss der Vipera berus. Ztschr. f. Chir. v. Chir., Osterode, 1841, I, pt. 2, 36–39.

Recovery from the bite of the cobra de Capello. Lancet, Lond., 1859, II, 13.

REDI (F.). Lettera sopra alcune opposizioni fatte alle sue osservazioni intorno alle vipere. Scritta alli signori Alessandro Moro e Abate Bourdelot. 31 pp. 4°. Firenze, 1685.

REDI (F.) Osservazioni intorno alle vipere. Eda lui scritte in una lettera all' illustrissimo signor Lorenzo Magalotti. 1 p. l., 91 pp. 4°. Firenze, 1664.

REIMONENSQ. Un jeune garçon de dix-neuf ans qui avait été mordu par un serpent. J. de méd. de Bordeaux, 1853, 748–750.

Remède contre la morsure de vipère. [Mém. Acad. r. d. sc. 1747.] Collect. acad. d. mem., etc., Par., 1785, X, 445.

REMEDIOS MONTEIRO (J.). Do permanganato de potassa contra o veneno das cobras. Gaz. med. da Bahia, 1881–2, 2. S., VI, 197–199.

REMER (W.). Folgen des Bisses einer giftiger Schlange. J. d. prakt. Heilk., Berl., 1814, XXXVIII, 1 St., 47–60.

RENGGER (J. R.). Ueber die Wirkung des Bisses der südamerikanischen Giftschlangen, und die von mir dagegen angewandte Heilmethode. Arch. f. Anat. u. Physiol., Leipz., 1829, 271–298. *Also, trans.:* J. compl. du dict. d. sc. méd., Par., 1830, XXXVII, 246–264.

Report of the Commission appointed to investigate the influence of artificial respiration, intravenous injection of ammonia, etc., in Indian and Australian snake poisoning. Indian Ann. M. Sc., Calcutta, 1875, XVII, 191–252, 4 pl.; with Appendix, Nos. 1–3, pp. I–XCVIII.

Report of the special committee on the subject of snake poisoning. Austral. M. J., Melbourne, 1877, XXI, 104, 151, 184.

Return showing the number of deaths from snake bites, in the year 1869, in the Province of Bengal. Population, including Orissa and

Assam, 48,358,134. Indian M. Gaz., Calcutta, 1870, V, Suppl., pp. 1–4.
Richards (V.). Experiments on snake poison. Indian Ann. M. Sc., Calcutta, 1870, No. XXVII, 177–202.
Richards (V.). Case of cobra bite. Indian M. Gaz., Calcutta, 1871, VI, 130.
Richards (V.). Treatment of snake poisoning by artificial respiration. Indian M. Gaz., Calcutta, 1872, VII, 247.
Richards (V.). Snake poisoning antidotes. The nature of snake poison, and its action on the blood. Indian Ann. M. Sc., Calcutta, 1872-3, XV, 163–176.
Richards (V.). Experiments with reputed antidotes and on artificial respiration in snake poisoning. Rep. on . . . san. improve. in India, 1872-3, Lond., 1873, 264–275.
Richards (V.). Dr. Fayrer's treatment of snake bite by artificial respiration. Indian M. Gaz., Calcutta, 1873, VIII, 118–120.
Richards (V.). The treatment of snake bite by intravenous injections of ammonia. Med. Times and Gaz., Lond., 1873, I, 639.
Richards (V.). Experiments with strychnine as an antidote to snake poison. Med. Times and Gaz., Lond., 1874, I, 595–597.
Richards (V.). Experiments with snake poison. Indian Ann. M. Sc., Calcutta, 1873–4, XVI, 285–301.
Richards (V.). Report on the snake bite cases which occurred in Bengal, Behar, Orissa, Assam, Cachar, etc., during the year 1873-4. Indian M. Gaz., Calcutta, 1876, XI, 96–100.
Richards (V.). A fatal case of snake bite, intravenous injection of ammonia; remarks on the application of the ligature in snake bite. Indian M. Gaz., Calcutta, 1876, XI, 320.
Richards (V.). A case of snake poisoning; no treatment; recovery. Lancet, Lond., 1878, I, 530.
Richards (V.). Snake poisoning and its treatment. Indian M. Gaz., Calcutta, 1880, XV, 309.
Richards (V.). A case of snake bite; hypodermic injection of permanganate of potash. Indian M. Gaz., Calcutta, 1882, XVII, 44.
Richards (V.). Further experiments with permanganate of potash, liq. potassæ, and iodine in cobra poisoning. Indian M. Gaz., Calcutta, 1882, XVII, 199–202.
Richards (V.). Notes on Dr. Wall's monograph on cobra and Daboia poisons. Indian M. Gaz., Calcutta, 1882, XVII, 239, 259.
Richards (V.). Permanganate of potash and liquor potassæ in snake poisoning. Lancet, Lond., 1882, I, 1097.
Richards (V.). Dr. Badaloni on the permanganate of potash. Lancet, Lond., 1883, II, 461.
Richardson (J. F.). Rattlesnake poison. Phila. M. Times, 1879, IX, 306.
Richter (M.). Vergiftung eines Mädchens durch den Biss der Coluber berus. Ztschr. d. deutsch. Chir.-Ver. f. Med., chir. u. Geburtsh., Magdeb., 1850, IV, 9–15.
Rivers (G. M.). The rattlesnake, its poison and antidote. South. M. Rec., Atlanta, 1874, IV, 505–513.
Robert. Morsure d'une vipère. Bull. gén. de thérap., etc., Par., 1834, VII, 307–309.
Robson (T.). Case of snake bite, in which bleeding was used as an auxiliary. Tr. M. and Phys. Soc., Calcutta, 1835, VII, 2. pt., 480–482.
Rogers (S.). Case of snake bite. Madras Q. M. J., 1839, I, 231.
Rose (W. G.). On snake bites. India J. M. and Phys. Sc., Calcutta, 1836, N. S., I, 448.
Rousseau (L. F. E.). Expériences faites avec le venin d'un serpent à sonnettes (Crotalus horridus.) J. hebd. de méd., Par., 1828, I, 291–296.
Roy (G. C.). Experiments with cobra poison. Indian M. Gaz., Calcutta, 1876, XI, 313.
Roy (G. C.). Remarks on the action of snake poison on the blood. Indian M. Gaz., Calcutta, 1877, XII, 315–317.
Roy (G. C.). A case of cobra bite, with remarks. Indian M. Gaz., Calcutta, 1882, XVII, 292–294.
Rufez. Enquête sur le serpent de la Martinique. France méd. et pharm., Par., 1860, VII, 81.
Salisbury (J. H.). Influence of the poison of the northern rattlesnake (Crotalus durissus) on plants. Proc. Am. Ass. Advance. Sc., 1851, Wash., 1852, VI, 336.

SALLE (O.). Untersuchungen über die Lymphapophysen von Schlangen und schlangenähnlichen Sauriern. Diss.-Inaug. [Göttingen.] 8°. Leipzig, 1880.

SALZER (L.). On the cobra poison. Calcutta J. M., 1868, I, 437; 1869, II, 19, 65, 140.

SCHAEFFER (E. M.). Interesting case of snake poisoning. Field and Forest, Wash., 1875, I, 12-14.

SCUNECK (J.). Is the bite of the heterodon, or spreading adder, venomous? Chicago M. J. and Exam., 1878, XXXVII, 585-587.

SCHORRENBERG (II.). Waarneming van eenen vergiftigen slangenbeet. Nederl. Lancet, Gravenh., 1849-50, 2. S., V, 249-258.

SCHULTZ (C.). A memoir on serpents. MSS. (written on one side), 4°. [n. p., n. d.]

SEBIZIUS (M.), Jn. Discursus med.-phil., de casu adolescentis cujusdam Argentoratensis mirabili; qui anno MDCXVII, octavo Aprilis, circa horam primam pomeridianam, mortuus in quodam paternarum aedium loco, adjacente ipsi serpente, à domesticis inventus fuit. 4°. Argentorati, 1617.

SEBIZIUS (M.). De casu adolescentis cujusdem Argentinensis mirabili, qui anno 1617, octavo Aprilis, circa horam primam pomeridianam, mortuus in quodam paternum aditum loco adjacte ipsi serpente, a domesticis inventus fuit. Nunc ob defectum exemplarium denuo recusus et appendice de quibusdam serpentum generibus. 4°. Argentorati, 1660.

SEIDEL. Ueber den Vipernbiss. Uebers. d. Arb., etc., der Schles. Gesellsch. 1847, Bresl., 1848, 229. Also: Ausz. a. d. Uebers. d. Arb u. Veränd. d. Schles. Gesellsch. f. vaterl. Kult., Bresl., 1847, 13. Also: Wehnschr. f. d. ges. Heilk., Berl., 1849, 77-79.

SEN (J. K.). A case of snake bite treated by injection of liquor ammoniae into the veins; death. Indian M. Gaz., Calcutta, 1874, IX, 43.

SEREINS. Morsure de vipère; traitement par des injections hypodermiques d'acide phrnique; guérison rapide. Union méd., Par., 1882, 3. S., XXXIII, 1062.

Serpents venimeux. Dict. encycl. d. sc. méd., Par., 1881, 3. S., IX, 387-417.

SHAPLEIGH (E. B.). Death from rattlesnake bite. Tr. Coll. Phys. Phila., 1868, N. S., IV, 263. Also: Am. J. M. Sc., Phila., 1869, April, 392.

SHORT (R. T.). Case of a lad, aged 17, who had been bitten by an average sized prairie rattlesnake. (Crotalus maroecayua.) Med. Arch., St. Louis, 1869, III, 564.

SHORTT (J.). Snake bite. Indian Ann. M. Sc., Calcutta, 1856, IV, 299-301.

SHORTT (J.). Experiments with the poison of the Cobra di Capello. Lancet, Lond., 1868, I, 556, 615.

SHORTT (J.). Experiments with snake poison. Madras Month. J. M. Sc., 1870, I, 214, 275.

SHORTT (J.). Experiments on the cobra snake poison in the theatre of Madras Med. College, January 29, 1870. Madras Month. J. M. Sc., 1870, I, 361-364.

SHORTT (J.). Cases of snake bite, etc. Madras Month. J. M. Sc., 1870, I, 351, 451

SHORTT (J.). Cases of snake bite continued Madras Month. J. M. Sc., 1870, II, 7, 101, 348, 422.

SHORTT (J.). Case of snake (cobra) bite, successfully treated by suction, liquor potassæ, and brandy. Lancet, Lond., 1870, I, 540-542.

SHORTT (J.). Antidote to the cobra poison. Madras Month. J. M. Sc., 1870, II, 249-252.

SHORTT (J.). Review of cases of snake bite. Madras Month. J. M. Sc., 1871, III, 81-91.

SHORTT (J.). Experiments with snake (cobra) poison, commenced at Combaconum, in the Tanjore District, on the 21st Sept. 1866. Madras Month. J. M. Sc., 1871, IV, 16, 165, 346, 426; 1872, V, 199, 286; 1872, VI, 17, 249.

SHORTT (J.). Cases of snake bite. Madras Month. J. M. Sc., 1872, V, 359-363.

SHORTT (J.). Snake poison, treated successfully with liquor potassæ; with remarks. Lancet, Lond., 1882, I, 725.

SHOWERS (C. I.) and PUNDIT (P. N.). Experiments on the action of snake bite poison and its antidote. Indian M. Gaz., Calcutta, 1869, IV, 1-3.

SHULER (W M) Snake bites Med Gaz.,
N Y., 1881, VIII, 292.

SIBERGUNDI Beobachtung eines Falles von
Vergiftung durch einen Schlangenbiss
Heidelb. klin. Ann , 1834, X, 392–400.

SIEVWRIGHT (F) Case of snake bite successfully treated Month J. M. Sc., Edinb.,
1842, II, 257.

Singular effect of snake poison. Cincin. Lancet
and Observ , 1860, N S., III, 318.

[SIRCAR (M. L.).] On the pathogenetic action
of the cobra poison [Cases.] Calcutta
J M., 1868, I, 121–129.

SMITH (C J). Snake poison and its antidote.
Brit. M J., Lond., 1868, I, 164.

SMITH (J. R). Iodine in snake bite. West.
Lancet, Cincin., 1850, XI, 284.

SMITH (T I.). Case of snake bite successfully
treated by vene-section. Tr. M. and Phys.
Soc. Calcutta, 1836, VIII, 95–100.

Snake (On) poison. Med Press and Circ.,
Lond., 1875, I, 334.

Soda an antidote for snake bite, and the sting of
poisonous insects St. Louis M. and S.
J., 1854, XII, 403.

Some account of the Tanjore antidotes for the
bite of a mad dog; and also for the bite
of venomous serpents. Lond. M. J., 1789,
X, 283–295.

SPALDING (I.). Geschichte der Einführung und
des Gebrauches der Scutellaria lateriflora,
als eines Mittels zur Verhütung und Heilung der durch den Biss wüthender Thiere
veranlassten Wasserscheu. [Transl. from
the English by Dr Heusinger.] Mag. f.
d. ges. Heilk., Berl., 1821, IX, 80–114.

SPINDLER (ALBERT RICHARD CONSTANT). Disquisitio physiologico-pathologica circa
virus viperarum hujusque effectuum nociferorum therapia. Diss. inaug 27 pp.
4°, Jenæ, ex off. Mauki (1823).

STAHMANN. Ueber Viperubiss. Ztschr. f. Med.,
Chir. u. Geburtsh., Magdeb., 1857, XI,
435–442.

STANLEY (E.). Bite of a rattlesnake. Buffalo
M. J. and Month. Rev M. and S. Sc.,
1853-4, IX, 464.

STAPLES (G. M.). Case of poisoning by a rattlesnake bite treated with bi-sulphite of soda.
Med. and. Surg. Reporter, Phila., 1865,
XIII, 279.

STEINDACHNER (F.) Herpetologische Notizen.
8° [Wien, 1867.] Repr. from: Sitzungsb. d. k. Akad d. Wissensch., Wien,
1867, LV, I, Abth Febr.

ST. JOHN (J. H.). On the poison of the rattlesnake. Lond. Med. Reposit., 1827,
XXVIII, 445.

STOCKBRIDGE (W.). Snakes and snake bites at
the South-West. Boston M. and S. J.,
1843, XXIX, 40–43.

SUTHERLAND (G. S.). Two fatal cases, believed
to have resulted from snake bite at Mount
Abu, Rajputana, India, Lancet, Lond.,
1880, II, 129.

TARA (A.). Caso di avvelenamento per morso
viperino. Gazz. med. ital. lomb., Milano,
1858, 4. S., III. 309.

TAYLOR (A. S.). Action of the poison of the
cobra di Capello, or *Naja tripudians.*
Guy's Hosp. Rep., Lond., 1874, 3. S.,
XIX, 297–309.

TAYNTON and WILLIAMS. Bite of the viper.
Lond. M Gaz., 1833, XII, 464.

THACHER (J.). [A case where a rattlesnake
struck its fangs into a man's hand; treatment; recovery.] Boston M. Intellig.,
1823, I, 62.

THIÉBAUT. Observation d'un cas de morsure de
vipère ayant causé des accidents trèsgraves, rapidement améliorés par la cautérisation au fer rouge. Ann. Soc. méd. de
l'arrond. de Neufchateau, 1874, I, 16-18.
Also: Rev. méd. de l'est, Nancy, 1876,
V, 245–247.

THIERRY DE MAUGRAS. Notice sur la vipère à
cornes. Rec. de mém. de méd., chir. . . .
mili., Par., 1847, 2. S., III, 324–336, 1 pl.

THOMAS (J. R). Snake bites treated by brandy.
Northwest. M. and S. J., Chicago, 1855,
XII, 395.

Three cases of snake bite (Phoorsa). Tr. M. and
Phys. Soc Bombay (1849–50), 1851, No.
X, App., 308–311.

TIEGEL (E.). Notizen über Schlangenblut. Arch.
f. de ges. Physiol., Bonn, 1880, XXIII,
278–282.

TISSEIRE (L. T.). De la vipère cornue (bicorne),
du Sud de l'Algérie. Gaz. méd. de
l'Algérie, Alger, 1858, III, 11, 41, 86, 104.

TIXIER (V.). Morsure de serpent à sonnettes;
médecine des osages. Gaz. d. hôp., Par.,
1841, 171.

TIXIER (V.). Rapport sur les vipères. Soc. d. sc. méd de Gannat, Compt. rend., 1860–1, XV, 60–66

TIXIER (V.) Morsures de vipères. Soc. d. sc. méd de Gannat, Compt. rend., 1872, XXVI, 53–60.

Tödtlicher Otternbiss. Mag. f. d. ges Heilk., Berl., 1826, XXI, 545.

TOMES (C. S.). On the structure and development of the teeth of ophidia Phil. Tr. Lond., 1875, CLXV, pt. 1, 297–302, 1 pl.

TOMES (C. S.). On the development and succession of the poison fangs of snakes. Phil. Tr., 1876, Lond., 1877, CLXVI, pt. 2, 377–385, 1 pl.

TRACY (J. G.). On the uvularia grandiflora, as a remedy for the bite of the rattlesnake. N. York M. and Phys. J., 1828, VII, 65–68.

TRAVERS (B.). On the arsenical treatment of cases of snake bite. Assoc. M J., Lond., 1853, 812.

TRESTRAIL (J. C.). Case of snake bite. Tr. M. and Phys. Soc Bombay, 1855, N. S., II, 303.

VON TSCHUDI (J J.). Bemerkungen über den Biss der Giftschlangen. Wien. med. Wchnschr., 1853, 470.

VON TSCHUDI (J J.). Eine Vergiftung durch eine Klapperschlange. Wien. med. Wchnschr., 1865, 371–373.

TURNER (G.). Bite of the adder. Lond. M. Gaz., 1837, XX, 472.

VALDES (J. L.). Metodo sencillo para descubrir el envenenamiento por el cobre. Repert. Med.-Habanero, Habana, 1843, 2. S., 123

VALENTIN (G.). Einige Beobachtungen über die Wirkungen des Viperngiftes Ztschr. f. Biol., München, 1877, XIII, 80–117.

VIAUD-GRAND-MARAIS (A.). Études médicales sur les serpents de la Vendée et de la Loire-Inférieure. J. de la sect. de méd. Soc. acad. de la Loire-Inf., Nantes, 1860, N. S., XXXVI, 93, 209.

VIAUD-GRAND-MARAIS (A.). Du venin de la vipère. Gaz. d. hôp., Par., 1867, 365, 369.

VIAUD-GRAND-MARAIS (A.). De la léthalité de la morsure des vipères. Gaz. d. hôp., Par., 1868, 245, 258.

VIAUD-GRAND-MARAIS (A.) Description de la maladie produite par l'inoculation du venin de la vipère. Gaz d hôp., Par., 1869, 190, 210.

VIAUD-GRAND-MARAIS (A). Du traitement à suivre dans le cas des morsures de vipères. Monit. scient., Par., 1871, 3. S., I, 371–383.

VIAUD-GRAND-MARAIS (A.). Quelques plantes américaines employées contre les morsures des serpents venimeux. Rev méd. franç. et étrang., Par., 1874, I, 362–371

VIAUD-GRAND-MARAIS (A.). De la léthalité de la morsure des vipères indigènes. Ass. franç. p. l'avance. d. sc. Compt.-rend. 1875, Par, 1876, IV, 1059–1064.

VIAUD-GRAND-MARAIS (A.). Note sur le vichamaroundon, les pilules de Tanjore, les pierres à serpents et quelques végétaux employés dans les Indes contre les morsures envenimées. J. de méd. de l'ouest, Nantes, 1879, 2. S., III, 30–40.

VIAUD-GRAND-MARAIS (A.) Noticia sobre el vichamarundu, las pildoras de Tánjore, las piedras de serpientes y algunas vegetales empleados en la India contra las mordeduras venenosas. Anfiteatro anat., Madrid, 1880, VIII, 72–74.

VIAUD-GRAND-MARAIS (A.) Note sur l'envenimation ophidienne étudiée dans les différents groupes de serpents. J. de méd. de l'ouest, Nantes, 1880, XIV, 34–55.

VIAUD-GRAND-MARAIS (A.). L'envenimation ophidienne étudiée dans les différents groupes de serpents. Gaz d. Hôp., Par., 1880, LIII, 901, 917, 942, 958, 990, 1029.

VIAUD-GRAND-MARAIS (A.). Serpents venimeux. Dict. encycl. d sc. méd., Par., 1881, 3. S., IX, 387–417.

VULPIAN (A.). Sur l'action du venin du cobra di Capello (Naja vulgaire, serpent à lunelles, serpent à coiffe). Arch., de physiol., Par., 1869, II, 123–126.

WAGNER. Otterbissvergiftung, durch äusserliche, einfache Behandlung beseitigt. Wissensch. Ann. d. ges Heilk., Berl., 1834, XXIX, 459–461.

WALL (A. J.). Report on the physiological effects of the poisons of the Naja tripudians and the Daboia Russellii. Rep. on san. meas. in India, 1876–7, Lond., 1878, 229–249.

WALL (A. J.). On the differences in the physiological effects produced by the poisons of certain species of Indian venomous snakes. Proc. Roy. Soc., Lond., 1881, XXXII, 333–362.

WALLIS (W.). Death from the bite of a viper. Brit. M. J., Lond., 1871, I, 560.

WARING (E. J.). Antidotes to snake bites. Madras Q. J. M. Sc., 1861, III, 336–353.

WARR (A. V.). Two cases of snake bite. N. Orl. M. and S. J., 1861, XVIII, 187.

WEBB (J. H.). Treatment of snake bite. Austral. M. J., Melbourne, 1872, XVII, 154–159.

WEBB (J. H.). The ammonia treatment. Austral. M. J., Melbourne, 1876, XXI, 159–165.

WERNEKINCK (F.). Erfahrung über die Wirkungen des Viperubisses in Westphalen. (2 cases.) Abhandl. d. ärztl. Gesellsch. zu Münster, 1829, I, 84–96.

WESTON (P.). On the poison of the common adder. Lancet, Lond., 1859, I, 529.

WHITE. A case of cobra poisoning treated by incision, and liquor potassæ locally and internally, terminating fatally in one hour and twenty minutes. Med. Times and Gaz., Lond., 1873, II, 413.

WHITING (L. E.). Bite of a rattlesnake; recovery. Boston M. and S. J., 1850, L, 258.

WHITMIRE (J. S.). Iodine an antidote to the venom of the rattlesnake, Crotalus, and in the treatment of the bite. Chicago M. J., 1860, N. S., III, 267–270.

WILLIAMS (J.). On the cure of persons bitten by snakes. Pacific M. and S. J., San Fran., 1858, I, 96–100.

WILLIAMS (S. W.). Rattlesnakes—*Crotalus horridus*. Boston M. and S. J., 1847, XXXVII, 449–453.

WILLIAMSON (F.). Poisoning by the bite of a rattlesnake. Boston M. and S. J., 1869, IV, 398.

WILSON (W. J.). Rattlesnake bite. Phila. M. Times, 1874–5, V, 183.

WIRT (W. H.). A case of serpent bite. Med. and Surg. Reporter, Phila., 1871, XXV, 112.

WOLFF. Folgen von Schlangenbissen. Arch. f. med. Erfahr., Berl., 1821, II, 42–44.

WOODS (P. H.). Five cases of snake bite. N. South Wales M. Gaz., Sydney, 1873–4, IV, 129–132, 1 pl.

WOODWARD (B.). Iodine as a remedy in rattlesnake bite. Northwest. M. and S. J., Chicago, 1856, XIII, 61.

WRIGHT (J. R.). Case of snake bite. Indian M. Gaz., Calcutta, 1873, VIII, 15.

WUCHERER (O.). Sobre a mordedura das cobras venenosas e seu tratamento. Gaz. med. da Bahia, 1867, I, 229, 241.

WUTH (E. M.). A suggestion as to the modern treatment of snake poisoning. Austral. M. J., Melbourne, 1883, N. S., V, 60–64.

DESCRIPTION OF PLATES.

PLATE I.

Fig. 1.—Poisoning by venom peptone of *Crotalus adamanteus*. Local appearances on section after death.
Fig. 2.—Poisoning by venom peptone of *Crotalus adamanteus*. Local appearances after death and before laying the part open. The œdematous prominent swelling is well shown, but is made rather too darkly red.
Fig. 3.—Local effects of venom peptone, when the poisoning is chronic. The grayish semi-gangrenous muscles are shown in contrast with the uninjured muscle of the opposite side.

PLATE II.

Fig. 1.—Extensive local lesions after death from venom globulin (dialysis globulin). From *Crotalus Adamanteus* venom.
Fig. 2.—Local lesions after death from a solution of dry Cobra venom—rabbit.
Fig. 3.—Contrast with Fig. 2 the profound local change caused in a rabbit by venom of Crotalus. In both cases fatal results took place in two hours, the doses having been small.

PLATE III.

Figs. 1 and 2 exhibit the increased adhesiveness of human-blood globules when the fresh blood has been mixed with fresh venom.
Fig. 3.—Naked-eye view of loop of mesentery in a cat showing effects of local application of venom of Crotalus. Extensive hemorrhages separate the two peritoneal layers, and are seen to have oozed through them freely.
Fig. 4.—First microscopic appearances of hemorrhage from capillaries of mesentery of cat.
Figs. 5, 6, 7.—Successive stages of increasing loss of blood.

PLATE IV.

Microscopic appearances of human blood on being mixed with fresh venom. The alteration in form and the elasticity and adhesiveness are well shown in Fields 1, 2, and 3—Photographs by Dr. Geo. A. Piersol.

PLATE V.

Extensive hemorrhagic lesions in abdominal organs of etherized rabbit poisoned by intra-peritoneal injection of venom of *Crotalus adamanteus*.

PLATE 1.

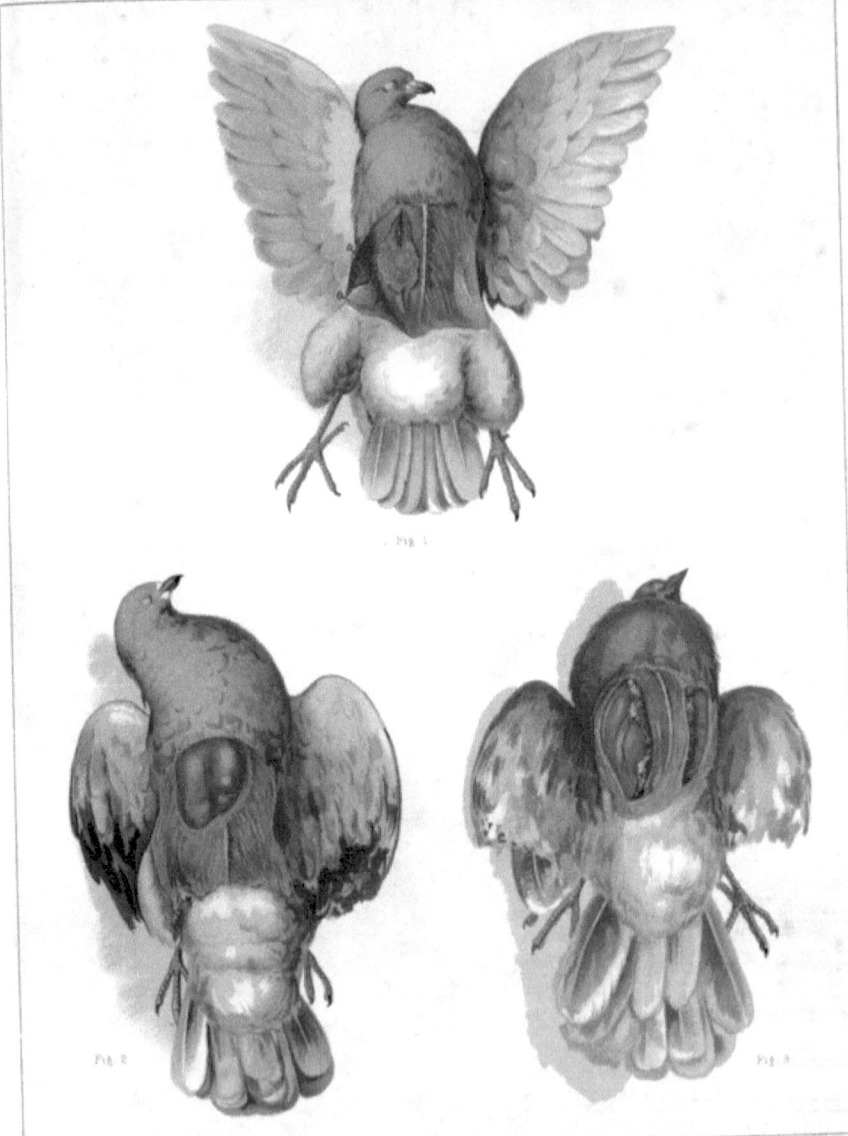

Fig. 1.

Fig. 2. Fig. 3.

PLATE II.

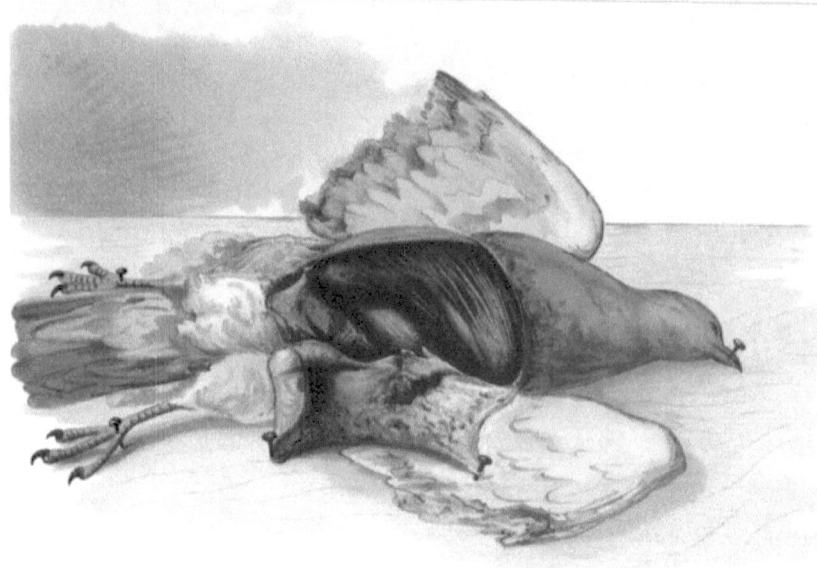

Fig 1.

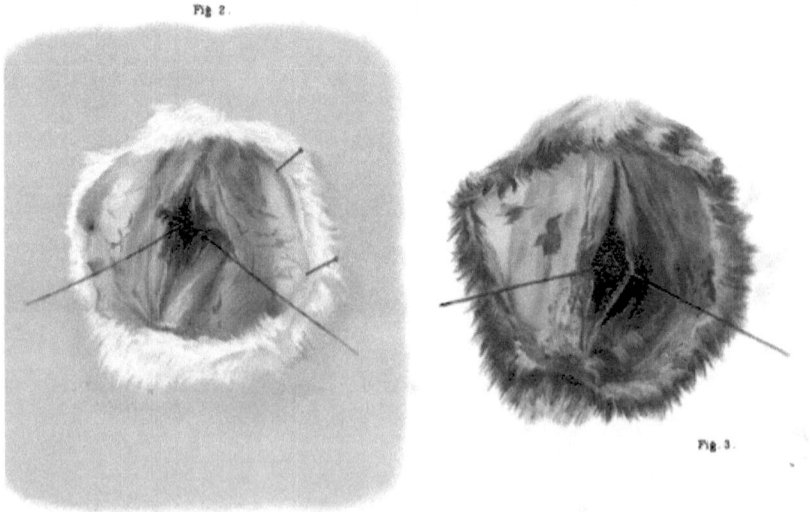

Fig 2.

Fig 3.

PLATE III.

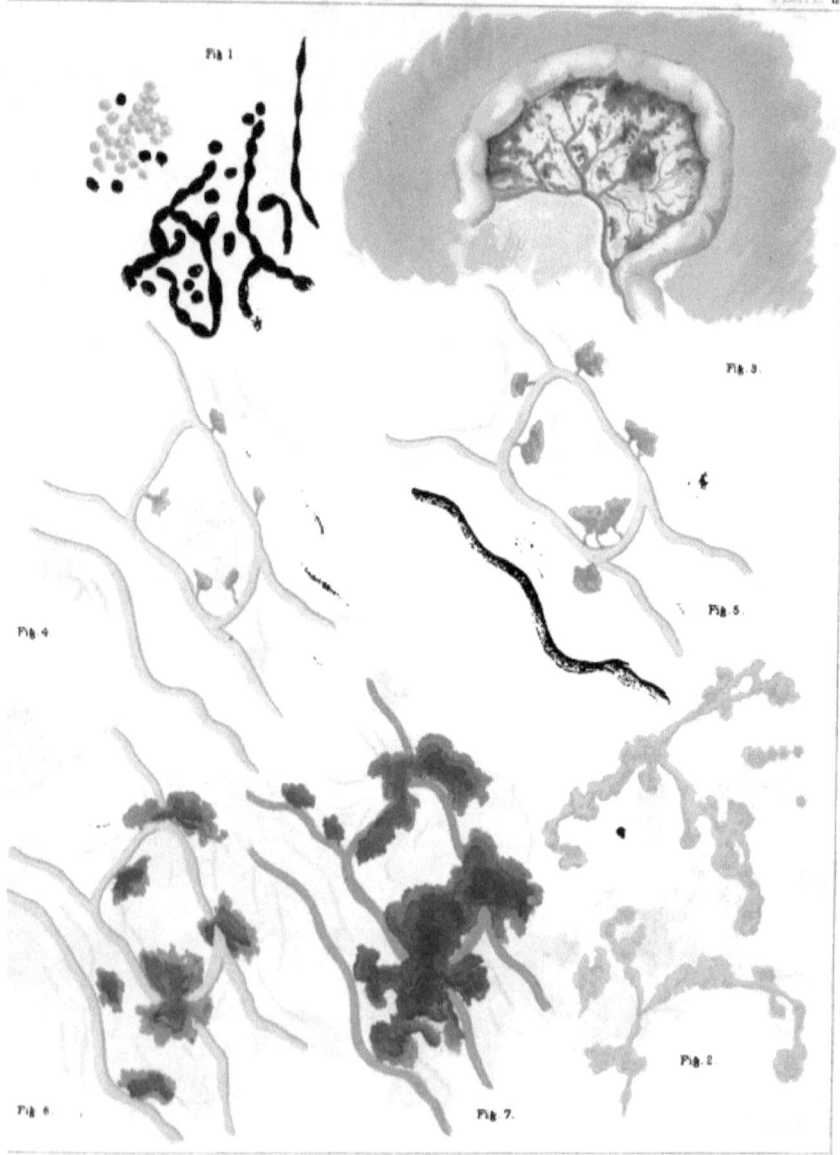

PLATE IV.

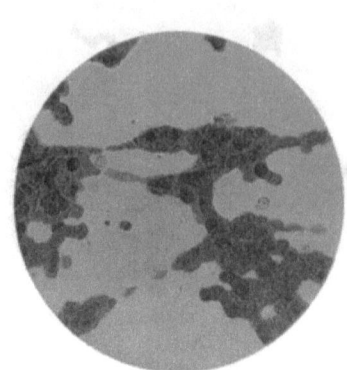

Fig. 1.

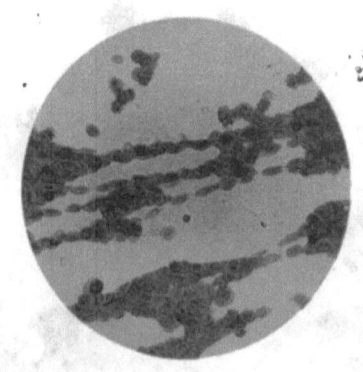

Fig. 2.

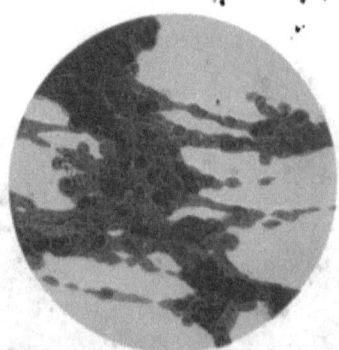

Fig. 3.

PLATE V

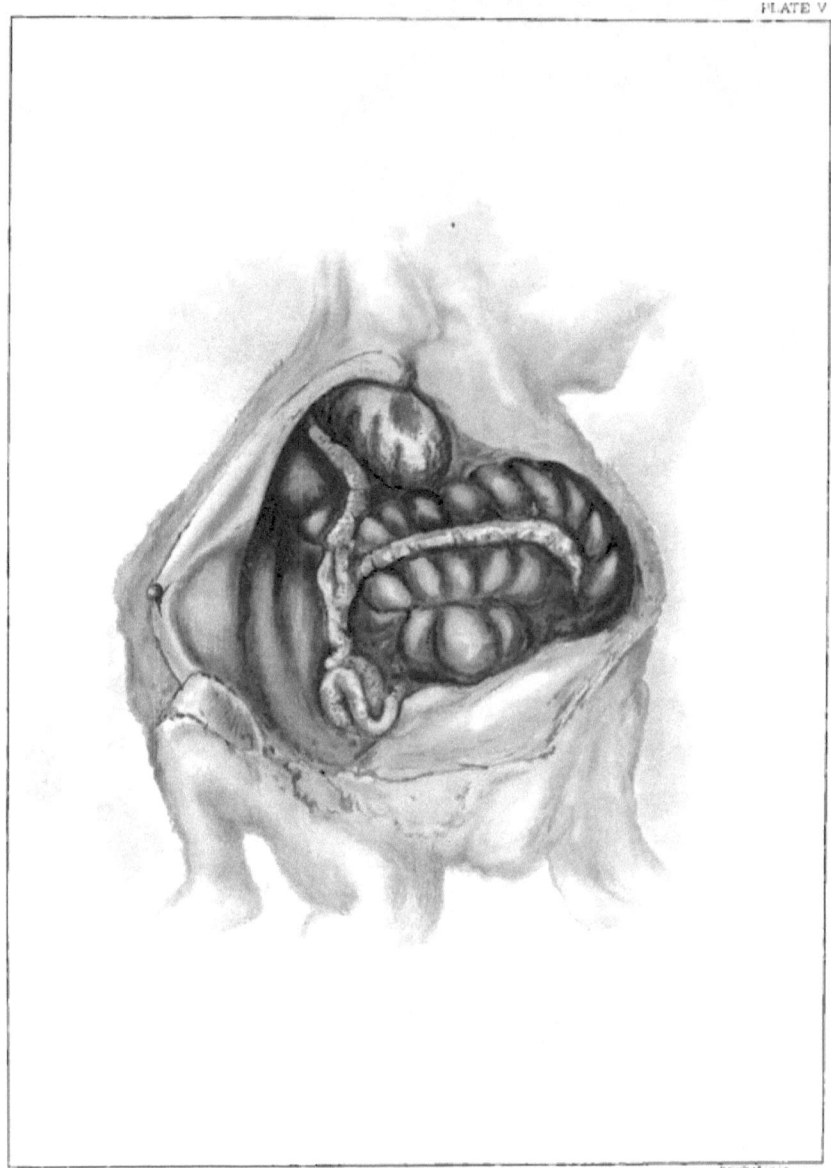

INDEX.

A.

Absorption of venom, 45, 154.
Acid, acetic, effect of, on toxicity of venom, 35.
Acid, bromohydric, effect of, on toxicity of venom, 37, 43.
Acid, hydrobromic, effect of, on toxicity of venom, 35, 43.
Acid, muriatic, effect of, on toxicity of venom, 34.
Acid, nitric, effect of, on toxicity of venom, 33.
Acid, sulphuric, effect of, on toxicity of venom, 34.
Acid, tannic, effect of, on toxicity of venom, 36.
Active principles of venom, 9, 157.
Activity of venom when applied to mucous surfaces, 44, 148; when applied to serous surfaces, 45, 149.
Age, effects of, on toxicity of venom, 21.
Agents, effects of various, on the toxicity of venom, 21.
Alcohol, absolute, effects of, on toxicity of venom, 29.
Alcohol, effects of, on toxicity of venom, 27.
Alkalies, effects of, on toxicity of venom, 29.
Alkaloids suspected in venom, 9.
Alum, effects of, on toxicity of venoms, 36.
Ammonia, effects of, on toxicity of venoms, 32.
Analysis of venoms—see chemistry of venoms.
Antidotes, local, 21, 43, 154, 157.
Appearances of venom when dried, 5.
Argentic nitrate, effects of, on toxicity of venoms, 39.
Arterial pressure, action of pure venoms upon, 85-102.
Arterial pressure, action of venom globulins upon, 102-112.
Arterial pressure, action of venom peptones upon, 112-118.
Artificial digestion, effect of, on toxicity of venoms, 42.

B.

Bacteria in venom, 6, 7, 133, 135, 136, 154.
Bile, effect of, on toxicity of venom, 42.
Blood clot, peculiarities of, caused by venom, 142.
Blood corpuscles, action of fresh venom upon, 143, 155.
Blood corpuscles, disintegration of, by venom, 141, 145.
Blood, effect of venom poisoning on, 138, 139, 140, 141.
Blood, extravasations of, 138-140, 146, 148, 154.
Bloodvessel walls, effect of venom upon, 146, 150, 154.
Boiled solution of venom, difficulties in filtering clear, 13, 17.
Boiling, effect of, on venom, 17, 26, 46, 47, 51.
Bouillon, putrefaction experiments with, 134.
Brain, effect of venom upon, 148.
Bromine, effect of, on toxicity of venom, 37.
Bromohydric acid, effect of, on toxicity of venom, 37.

C.

Care of serpents, 2.
Caustic alkalies, effects of, on toxicity of venoms, 29.
Caustic potash, effects of, on toxicity of venoms, 29.
Chemistry of venoms, 9, 153.
Ciliary motion, action of venom upon, 149.
Clot, blood, peculiar characters of, 142.
Coagulation of blood, action of venom globulin upon, 142.
Coagulation of blood, action of venom peptone upon, 142.
Coagulation of blood, action of venom upon, 138-141.
Color of venom, 5.
Coloring matter of venom, how separated, 7, 8.

(183)

Comparative local effects of different venoms, 55, 157.
Copper-venom-globulins, action on pulse-rate, 69-79.
Copper-venom-globulins, action on arterial pressure, 102-112.
Copper-venom-globulins, action on respiration, 125-130.
Copper-venom-globulins, comparative reactions of, 19.
Copper-venom-globulins, local actions of, 54.
Copper-venom-globulins, method of preparation, 11, 153.
Copper-venom-globulins, proportion in venom, 20.
Copper-venom-globulins, reactions of, 12, 15.
Cornea, the action of venom upon, 148.
Culture experiments with venom, 136.

D.

Daboia venom, 19, 55, 65, 92.
Death, cause of, in venom poisoning, 156.
Death, time of the occurrence of, 139-140.
Desiccation of venom, effects of, on toxicity of venom, 21.
Dialysis-venom-globulin, action of, on pulse-rate, 69-79.
Dialysis-venom-globulin, action of, on arterial pressure, 102-112.
Dialysis-venom-globulin, action of, on respiration, 125-130.
Dialysis-venom-globulin, local action of, 54.
Dialysis-venom-globulin, method of preparation, 12.
Dialysis-venom-globulin, proportion in venom, 20.
Dialysis-venom-globulin, reactions of, 12, 13, 15.
Dialyzed iron, effects of, on toxicity of venom, 40.
Difficulties attending researches with venom, 1.
Digestion of venoms, effects of, on toxicity of venoms, 22.
Dry heat, effects of, on toxicity of venoms, 22.
Drying venoms, loss in, 8.

E.

Ecchymoses—see hemorrhages.
Epithelium in venom, 6, 7.

F.

Ferric chloride, effects of, on the toxicity of venom, 40.
Ferrous sulphate, effects of, on the toxicity of venom, 39.

Filtration experiments with fresh venom to study morphological constituents, 133.
Filtration of venom through various substances, effects on the toxicity of venom, 42.

G.

Globulins—see venom globulins.
Granular matter in muscular tissue produced by venom, 141, 146.
Granular matter in venoms, 6, 133.

H.

Hemorrhages from venom poisoning, 139, 140, 146, 148, 154.
Hemorrhages, mechanism of, 149-152, 155.
Heat, effects of, on toxicity of venom, 35.
Heated venom, experiments with, 134.
Hydrobromic acid, effects of, on toxicity of venom, 35.

I.

Insoluble precipitate in venom, 9.
Insoluble precipitate in venom, general characters of, 10, 154.
Iodine and potassic iodide, effects of, on toxicity of venom, 37.
Iodine, effects of, on toxicity of venom, 37, 43.
Iron, liquor chloride of, effects of, on toxicity of venom, 41, 43, 154.
Iron, tincture of chloride of, effects of, on toxicity of venom, 41, 43, 154.

L.

Lesions from venom, macroscopical, 51, 137, 139, 140, 154.
Lesions from venom, microscopical, 6, 133, 141, 145, 146.
Liquor ferri chlor., effect on the toxicity of venom, 41, 43, 154.
Loss in drying venom, 8.
Lungs, alterations in, by venom, 139-141, 147.

M.

Macroscopical appearances produced by venom, 51, 137, 139, 140, 154.
Medulla, alterations in, 148.
Mercuric chloride, effect of, upon toxicity of venom, 39.
Mesentery, action of venom upon, 149.

INDEX. 185

Micrococci in venom, 6, 136, 154.
Microscopical appearances produced by venom, 6, 133, 141, 145, 146.
Moist heat, effect of, on toxicity of venom, 22-27.
Morphological constituents of venom, 133.
Motion, ciliary, action of venom upon, 149.
Motor nerves, action of venom upon, 49.
Mucous surfaces, action of venom upon, 44, 148.
Muriatic acid, effects of, upon toxicity of venoms, 33.
Muscular tissues, dry atrophy of, 140.
Muscular tissues, putrefaction experiments on, 135.

N.

Necrotic changes caused by venom, 135, 154.
Nerves, motor, effects of venom upon, 49.
Nerves, sensory, effects of venom upon, 49.
Nervous system, effects of venom upon, 48, 155.
Neutralization of venom, result of, 9.

P.

Pathology of serpent venoms, 133-152.
Peptone—see venom peptone.
Peroxide of hydrogen, effect of, on toxicity of venom, 39.
Physical characteristics of venom, 5.
Plan of isolation of poisons, defects in, 153.
Plan of study, defects in, 153.
Poisoning, chronic, 140.
Poisoning, rapid, 137.
Potash, effects on toxicity of venom, 29.
Potassic iodide, effects on toxicity of venom, 38.
Potassic iodide and iodine, effects on toxicity of venom, 37.
Potassic permanganate, effects on toxicity of venom, 38, 43, 153.
Preservation of venom, 5.
Pressure, arterial or blood—see arterial pressure.
Proteid constituents of venoms, 19, 20.
Pulse-rate, action of pure venoms upon, 56-69, 155.
Pulse-rate, action of venom globulins upon, 69-79, 156.
Pulse-rate, action of venom peptones upon, 79-84, 156.

R.

Reaction of venoms, 9.
Reactions of copper-venom-globulins, 12, 15.
Reactions of dialysis-venom-globulins, 12, 13, 18.
Reactions of water-venom-globulins, 11, 14, 17, 18.

24   July, 1886.

Reactions of venom peptones, 10, 13, 15, 18.
Respiration, action of pure venoms upon, 119-125.
Respiration, action of venom globulins upon, 125-130, 156.
Respiration, action of venom peptones upon, 130-132, 156.

S.

Saliva, analogy of venoms to, 154.
Salts in venom, 20.
Sensory nerves, action of venom upon, 49.
Serpent loop, 2.
Serpents, care of, 2.
Serpents, feeding of, 2.
Serous surfaces, effects of venom on, 45, 149.
Silver nitrate, effect upon toxicity of venom, 39.
Snake bile, effect upon toxicity of venom, 42.
Snake loop, 2.
Sodic hydrate, effect upon toxicity of venom, 32.
Solids in venom, 5, 6, 133, 154.
Specific gravities of venoms, 8.
Spermatozoa, effects of venom upon, 149.
Spinal cord, effects of venom upon, 49.
Sulphuric acid, effects of, upon the toxicity of venoms, 34.

T.

Tannic acid, effects of, upon the toxicity of venoms, 36.
Tincture of the chloride of iron, effects of, upon the toxicity of venoms, 41, 43.
Toxic elements in venom, 154.

V.

Venom globulins, 10.
Venom globulins, action on pulse-rate, 69-79, 156.
Venom globulins, action on arterial pressure, 102-112, 156.
Venom globulins, action on respirations, 125-130, 156.
Venom globulins, comparative reactions of, 14, 16, 18, 19.
Venom globulins compared with venom peptones toxicologically, 51, 118, 142, 156.
Venom globulins, defects in method of preparation, 153.
Venom globulins, distinguishing features of various, 13, 14, 16, 18, 19.
Venom globulins, general characters of, 10.
Venom globulins, how held in solution in venoms, 12.

Venom globulins, how prepared, 10.
Venom globulins, local actions of, 53.
Venom globulins, physiological peculiarities of, 47, 142, 156.
Venom globulins, proportions in venoms, 20.
Venom globulins, reactions of, 10.
Venom globulins, toxicity affected by boiling, 55.
Venom globulins, toxicity affected by drying, 53.
Venom peptones, 10.
Venom peptones, action on pulse-rate, 79–84, 156.
Venom peptones, action on arterial pressure, 112–118, 156.
Venom peptones, action on respiration, 130–132, 156.
Venom peptones, comparative reactions of, 19.
Venom peptones compared with venom globulins toxicologically, 51, 118, 142, 156.
Venom peptones, general characters of, 10, 17, 19.
Venom peptones, how prepared, 10, 13.
Venom peptones, local actions of, 51.
Venom peptones, peculiar characters of, 11, 17.
Venom peptones, physiological peculiarities of, 47, 156.
Venom peptones, proportions of, in venom, 20.
Venom peptones, reactions of, 10, 13, 15, 18.
Voluntary motion, effect of venom upon, 49.

W.

Water-venom-globulins, 10, 14, 16.
Water-venom-globulins, action on the pulse-rate, 69–79.
Water-venom-globulins, action on the arterial pressure, 102–112.
Water-venom-globulins, action on the respirations, 125–130.
Water-venom-globulins, comparative reactions, 18.
Water-venom-globulins, how prepared, 10.
Water-venom-globulins, local action, 54.
Water-venom-globulins, proportion in venoms, 20.
Water-venom-globulins, reactions of, 11, 14, 17.

Y.

Yellow pigment of venom, 6, 7, 8.
Yellow pigment of venom, separation of, 7, 8.

www.ingramcontent.com/pod-product-compliance
Lightning Source LLC
Chambersburg PA
CBHW020910230426
43666CB00008B/1390